# Experiments in Electronics, Instrumentation, and Microcomputers

**F. James Holler**
University of Kentucky

**James P. Avery**
University of Illinois

**Stanley R. Crouch**
Michigan State University

**Christie G. Enke**
Michigan State University

**The Benjamin/Cummings Publishing Company, Inc.**
Menlo Park, California • Reading, Massachusetts
London • Amsterdam • Don Mills, Ontario • Sydney

ISBN 0-8053-6918-X
  CDEFGHIJ-AL-89876543

 The Benjamin/Cummings Publishing Company, Inc.
2727 Sand Hill Road
Menlo Park, California 94025

## PREFACE

For scientists and engineers in every field, the tools of the trade are increasingly electronic in nature. In modern instruments, analog and digital electronic techniques perform primary functions of data manipulation and interpretation and of instrument and process control. It is the goal of this book to help provide a practical experience in and a working knowledge of the important electronic techniques and devices that are so dramatically affecting our professional lives. For most of the people using this book, electronics is not their principal occupation, but rather an essential tool in their work in other areas of engineering, life science, or physical science. Therefore, our goal has been to develop an efficient and practical sequence of experiments that is, at the same time, interesting, relevant, and current. We feel very strongly that it is exactly for such "users" of electronics that a laboratory course such as this is so important. The effective use of electronics in science is a combination of knowledge, skill, and art. Therefore, for electronics, as for sailing and banjo-picking, much of the learning is in the doing.

## The Integrated Circuit Revolution

Tiny and inexpensive integrated circuits now perform many functions that only a few years ago required literally racks of circuitry to achieve. The miniaturization and economics thus introduced have revolutionized the kinds of electronics products available, the techniques of their construction, and the modes of their repair. It would be surprising if this sweeping revolution did not also affect the teaching/learning laboratory.

The functions of integrated circuits are easily studied and these functions are readily combined into relatively sophisticated measurement or control systems. The ability and necessity to study these advanced functions is happily counterbalanced by the decreased need to study the details of circuit designs that are now available in IC form. Thus IC's allow and encourage a shift away from the "devices first" approach of electronics instruction to one which is more "top-down."

The new labor-intensive economics of electronics construction also points toward a minimization of specialized teaching equipment because of its high cost relative to high-volume commercial gear of equivalent function. These factors have been of prime concern during our design of these experiments.

## Approach

The general sequence of these experiments has been under development for over five years. The meters, function generator, and oscilloscope are introduced in the first two units so they are available for experimentation from the start. The basics of both analog and digital techniques are developed early in the context of a survey of electronic forms of data encoding. This allows the study of op amp analog techniques to lead smoothly into digitally-controlled (switched) analog circuits and non-linear analog applications. A deeper study of time and frequency encoding precedes further study of logic, flip-flops, and counting. Three units of microcomputers and their applications complete the study. Our experience with this sequence is that the direct relevance of each unit is effective in establishing and maintaining the experimenter's interest.

The sequence of units in this book follows one-for-one the sequence of chapters in the text <u>Electronics and Instrumentation for Scientists</u> This makes these two books a convenient package for study. However, it is important to emphasize that the order of topics was developed first in the laboratory; only from its success there did the parallel text evolve. It is our experience that a topical sequence that works well in the lab can be used to produce a pedagogically sound text, but that the reverse is frequently not the case. The experiments in this book do not just illustrate some of the principles developed in the text, but rather they provide an experience with practical electronic techniques, their current implementations and limitations. When used together, it is not this book that supports the text, but rather the other way around; the study of the text and problems prepares the student for the final and most crucial step in the process--demonstration of the ability to use practical electronics in the laboratory.

<u>Format</u>

The experiments in this book are divided into units and the units into sections. Each unit begins with a summary and a list of the parts necessary to perform it. Each section begins with a statement of the aims and methods of that section. The instructions for each experiment are complete, that is, all the data are provided and no references to the text or other sources should be necessary. The instructions are also rather thorough, especially in the early units. This has been done to reduce the difficulties encountered by beginning experimenters. Instructions in later units depend more and more on the capabilities obtained by the successful completion of the earlier units. For example, in Unit 1 range-setting and connection instructions are given for making a voltage measurement with the digital multimeter. In Unit 2, the instruction is simply, "Use the DMM to measure the voltage..."

Scattered through the experiments are questions and requests for plots of experimental data. These questions are an important supplement to the basic workbook format of the experiments. It is in considering these questions that the experimenter integrates the laboratory observations with the concepts in the text and explains the correlation or considers further possible applications. It is in answering these questions that the student demonstrates his/her individualized conclusions to the experiments. Correct answers to the questions generally demonstrate a much deeper understanding of the experiments and the principles underlying them than just the completion of the experimental observations.

## Application

This book is useful in a structured laboratory course or for self-study. The fourteen units are intended to require approximately equal time in the experimental laboratory. Suitable pre-lab preparation for each unit would be to study the corresponding chapter in Electronics and Instrumentation for Scientists and to do at least selected problems. As with the text, the order of units is somewhat flexible. (See "Notes to Experimenters," following.) Depending on the pace of the course and the background of the students, the material in this book is suitable for courses from 14 to 30 weeks in length.

The experiments have been designed to be performed with standard test equipment and, so far as possible, with standard, off-the-shelf experimental gear. For a description of the laboratory materials recommended, see the following section entitled, "Notes to Experimenters."

## Acknowledgments

Since this project has spanned half a decade or so, many students and teaching assistants at Michigan State University, the University of Illinois, and the University of Kentucky have contributed to its present form. Particularly, we would like to thank R. K. Calhoun, of the University of Kentucky, and Brian Seiler, John Eklund, and Malcolm Warren of the University of Illinois, and Thomas Atkinson, Peter Aiello, Bruce Newcome and Margaret McFarland of Michigan State University. We are also grateful to Dr. Timothy A. Nieman of the University of Illinois for providing the smoothing experiment of Unit 14-9. The preparation of the manuscript on word processor was done by Mrs. June Smith at the University of Kentucky, much of the art was drawn by Mr. Brian Seiler at the University of Illinois, and the final paste-up was performed by Ms. Jo Kotarski at Michigan State University. The final pages were prepared under a grant from Benjamin/Cummings. Such a project cannot be completed without the support and forbearance of those we are close to, so our special appreciation goes also to our families and friends, colleagues and students.

F. J. Holler

J. P. Avery

S. R. Crouch

G. C. Enke

## NOTES TO EXPERIMENTERS

To cover the wide range of material in this book in a time appropriate for learning a new tool requires that the time in lab is as efficiently spent as possible. This section is a brief summary of procedures that we have found to contribute to the efficiency and effectiveness of the laboratory experience.

### Prelab Preparation

Before beginning a unit, the entire related chapter should be studied and the assigned problems completed. The order of experiments in a unit is often not the same as the order of topics in the corresponding chapter of the text. It is often possible to follow the experiment instructions without careful study of the text, but the understanding of the goals and procedures will be lacking and this leads to poor efficiency and susceptibility to errors. Also read through the experimental unit--particularly the summary at the beginning of the unit and the purpose statement at the beginning of each section. If these do not make sense, study again the relevant parts of the text.

### Good Laboratory Procedure

Good lab practice begins with an orderly work space. Components should be returned to their allotted storage space. Patch wires with kinked wire ends should be trimmed, restripped, and stored according to approximate length (1"-2", 2"-5", > 5").

Circuit wiring and alteration should generally be done with the power off. It is especially important that signal sources be off, too. Many integrated circuits can be destroyed if signals are left connected when the power to the circuit is off. Plugging the power cords of signal sources into the switched outlet on the power supply helps avoid this problem. Before applying power, consider whether the instructions were understood and implemented correctly and what the expected response will be.

In case the circuit does not perform correctly, use this occasion to develop trouble-shooting skills. Most often, the connections have been made incorrectly, sometimes the IC is damaged, and occasionally the power supply or test equipment is faulty. Use the meters and oscilloscope to track down the fault. If a replacement part gives the same symptoms, do not just replace it with a third; your circuit is probably destroying the component. Only a certain amount of time in the trouble-shooting mode is productive. After a while, get help.

### Laboratory Reports

The blanks for data should be filled in directly during the laboratory time. These data provide the information which is the basis for many of the numbered questions in the experiments. The answers to these questions and the requested graphs and computer print-outs should be put on separate sheets. These sheets and the experiment/data pages comprise the core of the report for each unit. Additional writing such as a description of a scientific application for the studied material or a solution to an assigned design problem may also be required.

## Lab Equipment Recommendations For Instructors

The laboratory should be set up as a number of complete work stations. Each station contains an oscilloscope, function generator, digital multimeter, power supply, patch board frame, and some prewired experimental patchboards called "job boards." All of this equipment is standard hardware available from competitive suppliers except the patchboard frame. This unit performs a central interconnection function between the power supply and test equipment and the experimenter's patchboards. This frame, which can be easily constructed according to instructions in the Appendix, is well-worth the trouble in the increased efficiency of the experiments. Both time and components are saved by having experimenter's patchboards pre-configured with particular combinations of IC's. These savings more than offset the cost of the additional patchboards. The cost of each station, with careful purchasing, can be just over a thousand dollars (not including the microcomputer or curve tracer).

The lab station is thus low-cost, but the wiring techniques it teaches are representative of good experimental design practice. There is an absolute minimum of specially designed teaching equipment which reduces reliance on a sole-source vendor and allows choice among the most cost-effective options for equipment.

It is very desirable to have students working at the lab stations individually though sharing a station is possible. There is some additional equipment and material that should be available in the laboratory, but which can be shared among the students. For instance, a curve tracer is required for students doing Unit 7 and students working on Units 12 to 14 use an AIM-65 microcomputer. Since it is possible to avoid having all students doing these units at once (see next section), each station does not need a curve tracer and microcomputer.

The AIM-65 microcomputer was chosen for the last three units on the basis of its capabilities and its high cost-effectiveness. Its built-in printer, standard keyboard, and BASIC interpreter, as well as its user I/O port give it all the essential functions at a relatively low price. Other microcomputers could be used with modifications to some of the programs and instructions.

All of the required equipment specifications and details on the patchboard frame, power supply, and job boards are given in the Appendix.

## Variations on the Sequence of Units

Significant flexibility in the sequence of units is provided by the fact that units from 8 on do not have all previous units as prerequisites. This is shown by the table of prerequisites, below.

| Unit | Prerequisite Units |
|------|--------------------|
| 8    | 1-6                |
| 9    | 1-6                |
| 10   | 1-6, 9             |
| 11   | 1-7, 9-10          |
| 12   | 1-4                |
| 13   | 1-6, 8-10, 12      |
| 14   | 1-6, 8-10, 12-13   |

The ability to skip or postpone units 7 and 11 allows a 9 or 10-unit course to include the digital logic unit (10) and the introduction to microcomputers (12). In some contexts, the device-detail level of Units 7 and 11 may not be of interest to the majority of students. The microcomputer can be introduced at any time after Unit 4. This allows Unit 12 to be performed with many fewer microcomputers than there are work stations. Units on BASIC and/or ASSEMBLER programming can be inserted after Unit 12.

# TABLE OF CONTENTS

# UNIT 1.  ELECTRICALLY ENCODED INFORMATION

This unit introduces electrical signals in several data domains and the instruments used for their measurement.  A breadboard and interconnection system is introduced and used with a digital multimeter (DMM) in the measurement of voltage, current, and resistance.  Proper operation of the test equipment is introduced through a study of Kirchhoff's laws, the voltage divider and the current splitter.  The concepts of meter resistance and loading, including loading a voltage divider are explored.  Digital domain signals are generated by means of binary signal sources and detected by logic level indicators.

## Equipment

1. DMM with clip probes
2. DMM operations guide
3. Breadboard frame and power supply
4. Utility job board
5. Reference (REF) job board
6. Binary source and indicator (BSI) job board
7. Logic probe
8. Logic probe guide

## Parts

1. 9V transistor battery
2. Carbon and film resistors, various values, 1%, 5%, and 10% including 47 $\Omega$, 150 $\Omega$, 220 $\Omega$, 300 $\Omega$, 2-1 k$\Omega$, 10 k$\Omega$(5%), 12 k$\Omega$(5%), 15 k$\Omega$(5%), 220 k$\Omega$, 1 M$\Omega$, 10 M$\Omega$
3. Unknown resistors
4. Patch wires

## 1-1  DC VOLTAGE MEASUREMENTS WITH THE DMM

Objectives:  To become familiar with DMM operations for dc voltage measurements by the proper interpretation of the polarity indication, voltage range setting, and decimal point position; by the observation of the response of the ohmmeter to overrange conditions; and by the observation of the DMM response to open circuit and short circuit inputs.

Locate the digital multimeter (DMM), and examine its function, range, and on/off switches.  Note that there may be several different connectors for the various measurements functions (voltage, current, resistance).  Obtain from the instructor an operations guide for your specific DMM.  What should be the position of the function switch and the range switch in order to measure a 10.2 V dc signal without overranging the meter?  Where should the leads be connected to the DMM for the same measurement?

Function switch    _____

Range switch       _____

Lead connections _____

   Obtain a 9-V battery.  Set the DMM for dc volts and the range appropriate
for highest resolution (most sensitive scale without overrange).  Connect the
battery to the DMM with the probe clip leads (red meter terminal and red lead
to the positive battery terminal) and note the reading.  Reverse the leads and
again note the reading.

   V(battery) = _____          V(Battery, reversed) = _____

Note the DMM polarity indication with respect to that marked on the battery.
The polarity indication is + when the red meter terminal is connected to the
more (positive, negative) voltage.

   Switch the meter to ranges of higher and lower sensitivity and note the
result on the display.  If your DMM does not have automatic positioning of the
decimal point, take special note of the multiplier convention on the range
switch (X1, X10, etc.).
   Set the range switch to the most sensitive voltage scale.  Record the
reading obtained when the test leads are unconnected and when they are shorted
together.

   Unconnected reading or behavior _____

   Shorted reading _____

Question 1.  Suggest a reason for the difference in the two readings.


1-2  RESISTANCE MEASUREMENTS WITH THE DMM

Objectives:  To become familiar with DMM resistance measurements and resistor
             color codes by the observation of the response of the ohmmeter to
             open circuit and short circuit inputs, by the measurement of
             several resistances at maximum resolution, and by the comparison
             of measured and nominal (marked) values.

   Select the resistance mode with the function switch on the DMM.  If
necessary move one of the DMM input leads to the "Ohms" ($\Omega$) position.  Short
the leads of the DMM ohmmeter together and note the reading on each resistance
range (include units).

| Range | Reading, leads shorted |
|-------|------------------------|
|       |                        |
|       |                        |
|       |                        |
|       |                        |
|       |                        |

If you note a non-zero resistance on the most sensitive scale, this is due to
the resistance of the meter lead assembly.

Now disconnect the leads and note the response of the ohmmeter to an open circuit input.

Open circuit response _____

What is the maximum resistance your ohmmeter can measure? _____

How does the resistance between the open input leads compare with the maximum measurable value? _____

Measure the resistance of several resistors in the range of 100 Ω to 1 MΩ and with tolerances of 10%, 5%, and 1%. Use clip probe leads and the ohmmeter range setting that gives highest resolution. Record the values in the table below.

What is the rated accuracy for your ohmmeter in the resistance ranges used? _____

Compare the measured values to the nominal resistance values as indicated by the color code given in table 1-1. Is the error consistent with the resistor tolerance as indicated? Also obtain an unknown resistor from the instructor and report its value.

| Nominal Resistance | Nominal Tolerance | Measured Resistance | % Error |
|---|---|---|---|
| _____ | _____ | _____ | _____ |
| _____ | _____ | _____ | _____ |
| _____ | _____ | _____ | _____ |
| _____ | _____ | _____ | _____ |
| _____ | _____ | _____ | _____ |
| _____ | _____ | _____ | _____ |
| _____ | _____ | _____ | _____ |
| _____ | _____ | _____ | _____ |

Table 1-1

| Color | Figure | Multiplier | Tolerance |
|---|---|---|---|
| Black | 0 | 1 | — |
| Brown | 1 | 10 | ±1% |
| Red | 2 | 100 | ±2% |
| Orange | 3 | 1000 | — |
| Yellow | 4 | $10^4$ | — |
| Green | 5 | $10^5$ | ±0.5% |
| Blue | 6 | $10^6$ | ±0.25% |
| Violet | 7 | — | ±0.1% |
| Gray | 8 | — | ±0.05% |
| White | 9 | — | — |
| Gold | — | 0.1 | ±5% |
| Silver | — | — | ±10% |
| No band | — | — | ±20% |

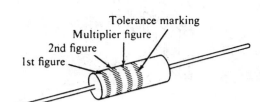

Tolerance marking
Multiplier figure
2nd figure
1st figure

Note:  Some 1% (precision resistors) possess five bands. For these, the third band represents an additional significant digit in the specified value, the fourth band is the multiplier to be applied to the three digits, and the fifth band represents a "quality" designation.

## 1-3  INTRODUCTION TO THE BREADBOARD SOCKET

Objectives:   To become familiar with solderless breadboard techniques through studying the contact arrangement on the breadboard socket and through the wiring and testing of series, parallel, and series-parallel combinations of resistors.

The basis of the breadboard socket is a spring metal clip with 5 pairs of opposing fingers as shown in figure 1-1a.  These metal clips are contained in partitioned sections of a plastic housing, one section of which is shown in figure 1-1b.  In the top of each plastic housing section, there is a row of 5 holes into which wires can be inserted for contact to the metal clip.  All wires plugged into the same metal clip are connected to each other.  The contact arrangement used in the breadboard socket is shown in figure 1-1c. This contains 64 rows of 5-contact groups along each side of the socket.  In addition, there are 4 long rows of connectors (5 groups of 5 contacts each) along the outer edges of each side.  The breadboard sockets are versatile because they accept and interconnect integrated circuits, most common wire terminal components, and wires.

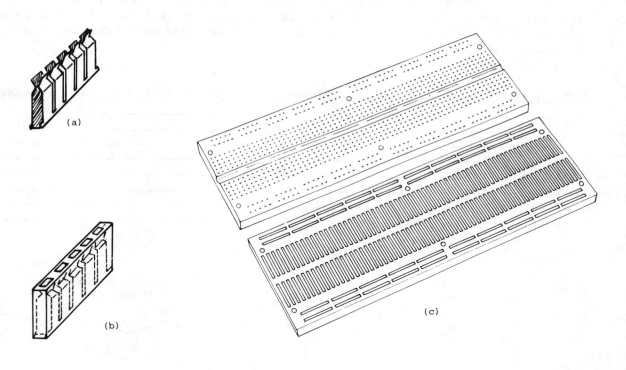

Fig. 1-1

In order to use the power supply and interconnection system recommended for these experiments, a printed circuit tab is inserted into the 5 pairs of contacts at one end of the socket as shown in figure 1-2.  The tab is secured to the socket by a pair of screws.  This entire assembly is referred to as a jobboard.  Do not separate the printed circuit tab from the breadboard.

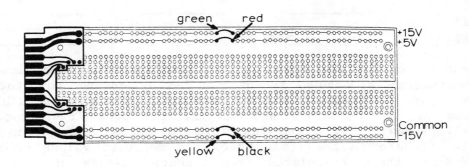

Fig. 1-2

Use a utility job board (one without prewired components) and the ohmmeter (with probe clips attached to short wires) to confirm the connection arrangement in the breadboard. Remove the short jumper wires on the edges of the job board. Confirm by resistance measurement that the contacts along the socket's edges are not connected across the middle gap. These contact strips are normally used as power busses. Reinstall the bus jumpers as shown in figure 1-2. Confirm that the power busses are now connected from end to end.

Locate three 5% resistors with values of 150 Ω, 220 Ω, and 300 Ω and insert them into the job board as shown in figure 1-3. By installing appropriate jumper wires, any desired interconnection (series, parallel, or series-parallel) of the three resistors can be obtained without removing the resistors from the socket.

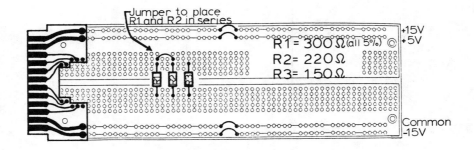

Fig. 1-3

Measure the resistances of each of the three resistors using the highest available resolution and record below. Use clip probe leads to connect the ohmmeter to the resistor leads.

|     | R(measured) | R(nominal) | % difference |
|-----|-------------|------------|--------------|
| R1  | _____ | _____ | _____ |
| R2  | _____ | _____ | _____ |
| R3  | _____ | _____ | _____ |

Connect two of the resistors together with a single patch wire to form a series combination as shown in figure 1-3. Measure the total series resistance. In a similar manner connect the three possible series combinations of two resistors and measure the total resistance of each combination. Connect all three resistors in series and obtain the total resistance. Compare in the table below the measured values of the series combinations to those <u>calculated</u> <u>from</u> <u>measured</u> <u>values</u> of the individual resistances.

|              | R(measured) | R(calc.)  | % difference |
|--------------|-------------|-----------|--------------|
| R1 + R2      | _____ | _____ | _____ |
| R2 + R3      | _____ | _____ | _____ |
| R1 + R3      | _____ | _____ | _____ |
| R1 + R2 + R3 | _____ | _____ | _____ |

Now disconnect the series jumpers and wire R1 and R2 in parallel. Measure the resulting resistance of the parallel combination. Repeat the measurement for all three resistors in parallel. In addition, wire the circuit shown in figure 1-4, and measure the resistance of the network. Include in the table a comparison with results calculated from measured values of the individual resistances.

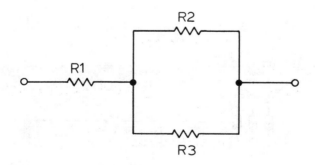

Fig. 1-4

|                | R(measured) | R(calc.)  | % difference |
|----------------|-------------|-----------|--------------|
| R1||R2         | _____ | _____ | _____ |
| R1||R2||R3     | _____ | _____ | _____ |
| R1 + (R2||R3)  | _____ | _____ | _____ |

## 1-4   INTRODUCTION TO THE BREADBOARD FRAME

Objectives:   To introduce the breadboard frame and become familiar with the power and signal interconnections provided by learning the proper insertion of the job boards, by confirming the connections between the job boards and the binding posts and BNC connectors, by observing the power distribution to the power busses of the job boards, and by illustrating the use of the system to connect a signal between a job board and an external test instrument.

Insert the utility job board into printed circuit (PC) card edge connections at the left-most position in the breadboard frame as illustrated in figure 1-5.

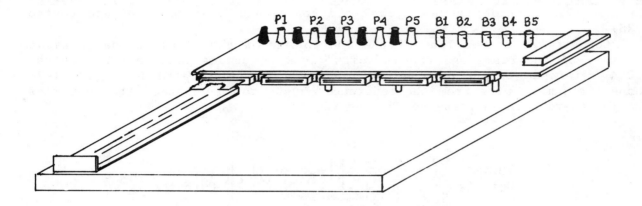

Fig. 1-5

All of the BNC connectors (a connector for shielded cable designed by Berkeley Nuclear Corp.) on the right and banana connectors on the left at the top of the frame are bussed to each of the job board positions via the PC edge connectors.  Refer to figure 1-6 and note the correspondence between the contacts on the top of the job board and the various input-output connectors at the top edge of the frame as shown in figure 1-5.  It should be noted that all of the black banana jacks, all of the BNC shields, the power supply ground return, the system common, and the chassis, are internally connected together. In electronics the color black is often used for connections to common.

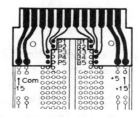

Fig. 1-6

Connect the ohmmeter between banana jack P1 and the corresponding wire socket on the job board and note the reading. Confirm the correspondence between the remaining top connectors and the job board contacts.

Change the DMM to the DC volts mode, connect the common lead to the system common on the job board, turn the power supply on, and measure the voltages on the power busses.

V(15) = _____    V(5) = _____

V(COM) = _____    V(-15) = _____

Shut off the main power and locate the reference (REF) job board. Install the REF job board in the frame, and turn on the main power. The main power should always be turned off while installing or removing job boards. The lower portion of the REF job board contains a voltage reference source (VRS).

Connect the common lead of the DMM to one of the black banana jacks at the rear of the frame and the + "volts" lead to banana jack P5. Locate the +10 V reference output on the REF job board as illustrated in figure 1-7. Connect a jumper wire from the reference output to the contact at the top of the job board corresponding to P5 (see figure 1-6).

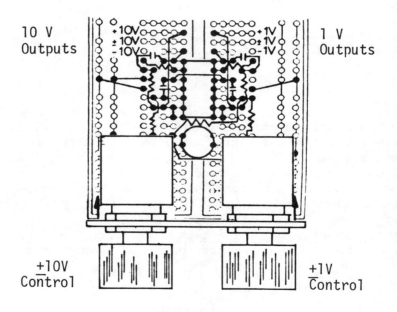

Fig. 1-7

Whenever possible, external devices should be connected to the job boards via the banana jacks or BNC connectors instead of using clips on the banana leads. This practice avoids congesting the patch wiring area and helps prevent accidental short circuits. It is therefore important for you to familiarize yourself with the connection scheme. Measure the voltage at P5 with the DMM and record the reading below.

V(ref) = _____

## 1-5  CURRENT MEASUREMENTS WITH THE DMM

Objectives: To become familiar with DMM operations for dc current measurements and to verify that the current is everywhere the same in a series circuit, by the connection of the ammeter into the circuit, in several different locations including one in which neither meter terminal is system common, and by the proper interpretation of the number displayed.

A schematic diagram of a simple series circuit is shown in figure 1-8a and b. It is not necessary to wire this circuit. In order to carry out a current measurement, it is necessary to open the circuit and to insert the DMM into the loop in series as shown in figure 1-8c. Turn off the main power, and wire on the utility job board the circuit of figure 1-8c. The pictorial of figure 1-8d is to aid you in this wiring. Note that the connection from the - side of the power supply to common has been made for you in the frame. DO NOT connect the +5 V power bus to the common bus. Doing so would short out the power supply.

Set the function switch of the DMM to the current mode, and adjust the range switch for lowest sensitivity (highest current). Attach the DMM leads to P3 and P4 via banana patch cords. At the DMM attach the leads according to the instructions given in your DMM operations guide. Turn on the main power and record the current in the loop. Reverse the DMM leads and note the effect.

I = _____        I(rev) = _____

By simply rearranging the jumpers on the job board, the current meter may be inserted in the position shown in figure 1-8e. Note that in this configuration, neither lead of the meter is attached to common. The meter is said to be floating. If one meter lead is accidently left connected to common, the floating connection will short out part of the circuit. Therefore, care should be taken when making floating connections.

I = _____

Now connect the circuit of figure 1-8f and again record the current from the DMM.

I = _____

Question 2. Explain what would happen if P3 were connected to system common in the circuit of figure 1-8e? In 1-8f? Do not attempt this measurement.

Question 3. Calculate the power dissipated in each resistor in the circuit of figure 1-8b. Show all work.

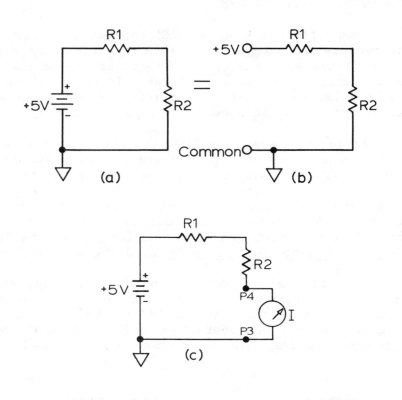

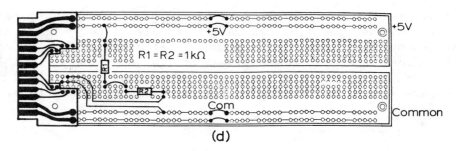

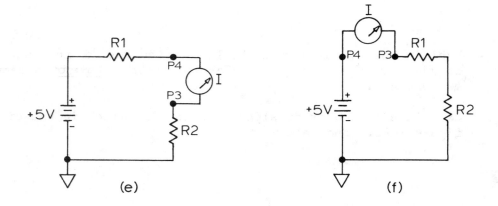

Fig. 1-8

## 1-6  VOLTAGE DIVIDER AND CURRENT SPLITTER

Objectives:  To demonstrate Kirchhoff's voltage  and current  laws and  their
application in voltage dividers and current splitters by wiring a
series combination of resistors and observing the voltage divider
operation, by observing the function of the continuously variable
voltage divider outputs of the VRS,  and by observing the current
splitting in a parallel resistor circuit.

### A.  Kirchhoff's Voltage Law

Measure precisely with the DMM the resistances of three 5% resistors with
nominal values of 10 kΩ (R1), 12 kΩ (R2),  and 15 kΩ (R3).   Now connect these
resistors in the series circuit shown in figure 1-9.   Apply power and measure
the voltage V(ad)  between points a and  d in the circuit.   Continue here the
practice of connecting the  DMM to the circuit through the  P terminals at the
rear of  the frame.   Also  measure the  voltage drops across  each individual
resistor.   Caution:  some  of these  measurements will  require a  floating
voltmeter (neither terminal to common).

R1 = _____        V(ad) = _____
R2 = _____        V(ab) = _____
R3 = _____        V(bc) = _____
                            V(cd) = _____

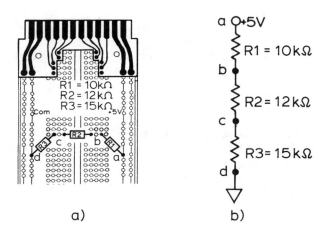

a)                                    b)

Fig. 1-9

<u>Question 4.</u>  Show that Kirchhoff's voltage law is obeyed.

### B.  Voltage Dividers

The circuit used in  the above experiment is a voltage  divider in that a
fraction of  the total  source voltage V(ad)  appears at the  points b  and c
(relative  to point d).   To obtain  the divider fractions  experimentally,
measure V(bd) and complete the table.

12

V(ad) = _____
V(bd) = _____
V(cd) = _____

Disconnect this circuit.

<u>Question 5</u>.   Calculate the theoretically expected  values for V(bd)/V(ad)  and
V(cd)/V(ad)  and compare with the measured ratios.  Use  the
measured rather than the nominal values of the resistors.

If  the  fixed  resistors  of  a  voltage  divider  are  replaced  by  a
potentiometer,  a variable voltage source is obtained.   The voltage reference
source job  board contains  two such  circuits to  produce ± 10 V  and ± 1 V
adjustable  outputs.    The   circuit  shown  schematically  in  figure 1-10
illustrates how  the potentiometer of  the VRS  is connected (prewired  on the
board) to produce the ± 10 V output.

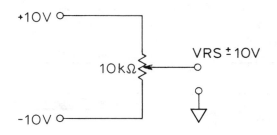

+10V

10kΩ          VRS ±10V

-10V

Fig. 1-10

Connect the DMM to the ± 10 V  VRS output and measure the range of output
voltages obtainable.

Range = _____

Adjust the VRS output to a value as near to 1.500 V as possible.  Use the most
sensitive DMM scale possible without  overranging.   Repeat  three  times,
approaching 1.500  V from  both directions.    To what  resolution is  the VRS
readily settable on the ± 10 V range?

V(out) = _____, _____, _____.

Resolution _____.

Repeat for the ± 1 V output.  Test the resolution at an output of 120.0 mV.

Range = _____

V(out) = _____, _____, _____.

Resolution _____.

<u>Question 6</u>.   Which of the VRS outputs should be used to obtain a 25 mV signal?
Justify your choice.

## C. Current Splitting and Kirchhoff's Current Law

Turn the power off and wire the circuit of figure 1-11 on the utility job board. Jumper the connectors to the DMM voltmeter so as to measure the voltage Va and turn the power on. Record the value of Va below. Now turn the power off, switch the DMM to the current mode, and insert it into the circuit in place of jumper W1, which connects the array to +5 V. Turn on the power and measure the total current I1 with the DMM set for maximum resolution (always start with the DMM on lowest sensitivity and increase). Compare I1 to the value expected from the measured Va and the known resistances.

Va = _____     I1 (measured) = _____

I1 (expected) = _____

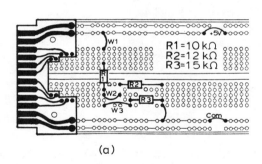

(a)

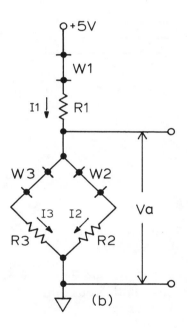

(b)

### Fig. 1-11

Replace W1, remove W2 and insert the current meter in place of W2. Measure I2 and record its value below. In a similar manner, measure I3.

I1 _____     I2 _____
I3 _____

<u>Question 7</u>. Show by appropriate calculations that Kirchhoff's current law is obeyed.

<u>Question 8</u>. Use the current splitting relationship to calculate I2 and I3 and compare to the measured values.

Disconnect all components from the utility job board, but leave the four power jumper wires in place.

## 1-7  METER RESISTANCE AND LOADING

Objectives:  To demonstrate the errors in voltage and current measurements due to the finite resistances of signal sources and measurement devices by observing a change in the voltage measured with the DMM due to a change in the source resistance, by determining the DMM input resistance through the measurement error, by estimating the amount of error in a voltage measurement, and by identifying the sources of loading error in current measurements with the DMM.

Connect a 1 MΩ (nominal) resistor on the utility job board.  Measure its resistance and record below.  Connect the circuit shown in figure 1-12.  Measure the voltages at point a and point b (relative to common) and record them below.

Rs = _____
Va = _____
Vb = _____

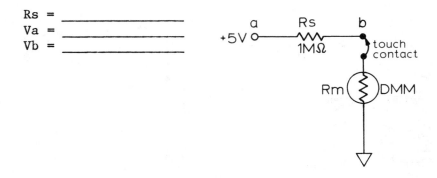

Fig. 1-12

Question 9.  From the values measured above, calculate the internal resistance of the DMM.

Question 10.  A 500 mV source has an internal resistance of 100 kΩ.  Show what the DMM would read for such a source.  Is the DMM reading incorrect?

Next a voltage divider output will be loaded with an external load.  The effect of the load on the divider output will be measured.  Measure the exact resistance of a 100 kΩ, a 220 kΩ and a 10 MΩ resistor.  Assemble the voltage divider of figure 1-13 on the utility job board.  Leave one end of the 10 MΩ resistor unconnected.

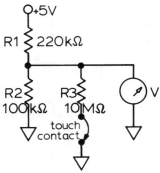

Fig. 1-13

Measure the voltage across the 100 kΩ resistor with and without the 10 MΩ resistor connected. Use the meter range which gives the greatest resolution.

R1 = _____          V(without) = _____
R2 = _____          V(with) = _____
R3 = _____

Disconnect this circuit.

Question 11.    Describe why the output voltage changes when the 10 MΩ load is connected.

Question 12.    From your measurements, calculate the divider output voltage when neither the voltmeter nor the 10 MΩ resistor is connected. Calculate the percentage change in the divider output voltage caused by using only the DMM to measure the divider output. Use the value for the DMM internal resistance calculated in question 9.

Now the internal resistance of the DMM in the current mode will be measured. Measure the resistances of a 1 kΩ and a 47 Ω resistor. Measure also the voltage of the +5 V supply. Connect the circuit of figure 1-14. Measure the current at maximum resolution with and without the 47 Ω resistor connected.

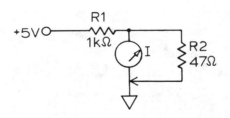

Fig. 1-14

R1 = _____          I(with) = _____
R2 = _____          I(without) = _____
5 V output = _____          (full scale) = _____

Question 13.    From the measured currents, calculate the internal resistance of the DMM in the current mode for the range selected. Assume that when R2 is connected, the total current through the meter and R2 is very nearly equal to the supply voltage divided by R1.

Repeat the measurement on a less sensitive scale.

I(with) = _____
I(without) = _____
I(full scale) = _____

Question 14. Calculate the current meter resistance for this less sensitive range. The DMM measures current by using a digital voltmeter (DVM) of fixed voltage range to measure the IR drop across an input resistor. This resistor (the input resistance) is selected by the range switch. What is the voltage range of the DVM used to measure the IR drop for current measurement in your DMM?

## 1-8 LOGIC LEVEL SOURCES AND INDICATORS

Objectives: To become familiar with binary or logic data encoding, its sources and indicators, by measuring the output voltage states of binary switch logic sources and the input voltage response of binary level indicators, and by generating signals representing the numerals 0-9 with four logic sources.

Insert the binary source and indicator (BSI) job board shown in figure 1-15 into the frame. Locate the outputs of the logic level sources. Measure the output voltages of source SA with the DMM for both states of the corresponding switch.

V(HI) = _____

V(LO) = _____

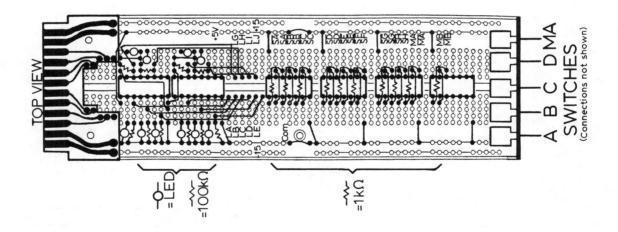

Fig. 1-15

The switch circuits of the switch register are <u>binary</u> <u>information</u> <u>generators</u> which give an electrical indication of the state of the switch. The light emitting diode (LED) circuits are <u>binary</u> <u>information</u> <u>detectors</u>. Thus, an ON LED indicates a HI logic level at its input. The use of switches and light states to indicate a binary number is shown in figure 1-16. Binary numbers are reviewed on page 5 of the text.

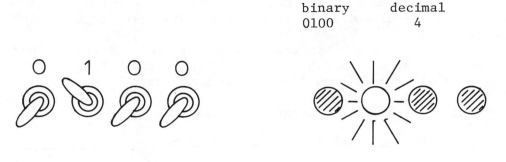

Fig. 1-16

Connect the output of a source to the input of an LED indicator. Toggle the switch and note the effect on the LED. Disconnect the switch output.

Connect 4 switch outputs to 4 LED indicators. Set the switches to produce the logic level outputs corresponding to the decimal numbers 0 through 9.

| Decimal No. | Binary Equivalent |
|---|---|
| 0 | _____ |
| 1 | _____ |
| 2 | _____ |
| 3 | _____ |
| 4 | _____ |
| 5 | _____ |
| 6 | _____ |
| 7 | _____ |
| 8 | _____ |
| 9 | _____ |

Disconnect the switches from the LEDs. Connect the output of a push button switch MA (see figure 1-15) to a LED indicator and the complementary output ($\overline{MA}$) to a different indicator. Confirm that the signal at MA is normally LO and is momentarily HI when MA is depressed. Confirm that the output $\overline{MA}$ is always in the opposite state to the output MA.

Adjust the VRS to ~+1.5 V. Keep the DMM connected to the VRS and also connect the VRS output to an LED indicator. Carefully increase the VRS output until the LED lights. Do not allow the VRS output to have a negative value or to exceed +5V. This can damage the indicator circuit. Record the transition threshold voltage for the LED indicator input.

V(thresh) = _____

Question 15. The LED indicator response interprets all input voltages which light the LED as HI and all that don't as LO. What are the ranges of HI and LO voltages as seen by the LED indicator?

Obtain a logic probe and connect the power leads to the frame power supply following the instructions in the manual provided. Touch the probe tip to a contact on the +5 V supply bus and note the probe LEDs. Now touch the probe tip to the ground bus and note the effect. Attach the probe to the output of a logic level source. Toggle the appropriate logic switch and note the probe response.

Touch the logic probe to the VRS output and carefully increase the voltage through a range from 0.5 V to 4.0 V. Note the effect on the indicator LEDs. Now decrease the voltage through the same range.

V(HI) threshold when input is increasing _____

V(LO) threshold when input is decreasing _____

<u>Question 16</u>. What are the ranges of HI and LO voltages as seen by the logic probe? In what ways might the logic probe be more convenient than the LED indicator?

# UNIT 2.   PERIODIC WAVEFORMS AND THE OSCILLOSCOPE

This set of experiments serves to introduce the oscilloscope (scope), the function generator (FG), and the ac scales of the digital multimeter (DMM). Effective operation of the oscilloscope is developed while it is used to characterize the signals available from the function generator. The response functions of low-pass and high-pass filters are determined and compared with the expected behavior. The frequency response functions of the scope ac inputs and DMM ac scales are interpreted in terms of the behavior of low-pass and high-pass filters. The DMM response to non-sinusoidal periodic signals is characterized.

<u>Equipment</u>

1. Basic station:
   dual trace oscilloscope (scope),
   DMM, function generator (FG),
   breadboard frame and power
   supply (PS), 2 banana leads for
   DMM, 4 BNC cables, patchwire
2. Manual for scope
3. Utility job board
4. Manual for FG

<u>Parts</u>

1. Resistors, 5%:  39 Ω, 270 Ω,
   1 kΩ, 10 kΩ, 100 kΩ
2. Capacitors:  0.001 µF,
   0.01 µF
3. Unknown filter

## 2-1   INTRODUCTION TO THE OSCILLOSCOPE AND FUNCTION GENERATOR

Objectives:   To become familiar with the waveform display function of the oscilloscope and the variety of waveforms available from the function generator by observation of the effects of the sensitivity and position controls of the scope, by connection of the FG output to the scope for observation of an FG output waveform, by determination of the repetition frequency of an observed signal through the scope sweep setting, and by the use of the second vertical input channel of the scope to observe two waveforms and the time relationship between them.

Turn the scope on with nothing connected to its inputs. Set the input switches of both channels to GND (see text page 50 for a discussion of ground) and the triggering mode to AUTO or NORMAL. If there is no input switch which can be set to GND on your scope, insert the utility job board in the breadboard frame, connect BNC cables to the scope input and to the frame, and connect the corresponding contacts on the utility job board to the common bus on the job board. Set the sweep rate to 1 ms/div (divisions on the scope graticule refer to the major divisions). Use the Ch 1 vertical position control to place the trace of Ch 1 along a graticule line in the upper half of the screen. Place the Ch 2 trace along a line in the lower half of the screen. The position of the trace with the input selector at GND is the zero reference point for vertical deflection. Observe the effects of the horizontal position, focus, and intensity controls. Describe the interaction, if any, between the focus and intensity controls.

Observations and description:

Connect the scope Ch 1 input to one of the breadboard frame connections. Insert the utility job board. Connect the +5 V supply voltage on the job board power bus to the scope input contact at the job board tab. Switch the Ch 1 input to DC and the sensitivity to 5 V/div. Record the vertical deflection of the trace for a 5 V signal.

_____ div for +5 V

Similarly measure and record the deflections produced by the +15 V and −15 V supplies in the frame.

_____ div for +15 V

_____ div for −15 V

Now readjust the vertical sensitivity and position controls so that the voltages of any of the three supplies can be checked without offscale deflection. There should be a difference of at least 5 divisions of deflection between the +15 V and −15 V deflections. Record the settings.

position of "0" line _____

vertical sensitivity _____

Connect the function generator (FG) signal output to one of the frame connections. Make sure you include a common connection. Connection to common is automatically made when a BNC connector is used to connect an external instrument to the frame or when the instrument common is connected to a black banana jack. On the job board, connect the FG output to the Ch 1 input of the scope. Select a 1000 Hz sine wave output on the FG. Set the scope trigger selector to Ch 1 (+) and the trigger input switch to AC. Set the Ch 1 vertical sensitivity to 2 V/div (calibrated). Change the Ch 1 input selector to DC. Make sure the scope time base is still set at 1 ms/div. Observe the waveform and the effects of the FG amplitude and offset controls over their entire range. Measure the maximum and minimum FG amplitude. Adjust the FG amplitude and offset controls to obtain a 5 V peak-to-peak signal centered on 0 V. How many complete cycles are observed in the scope display?

FG amplitude, maximum = _____ V(p-p)

Number of cycles displayed _____

Set the FG frequency so that there are exactly 4 divisions of horizontal deflection for each cycle (with the scope sweep rate still at 1 ms/div). Record the frequency setting on the FG dial and the frequency calculated from the observed horizontal deflection per cycle. Estimate the error in the frequency values above caused by uncertainty in reading the dial and display.

FG setting                _____ Hz

Measured from scope       _____ Hz

Estimated error, dial reading _____ Hz

Estimated error, display     _____ Hz.

Set the FG to 30 kHz and the oscilloscope time base (TB) to a value suitable for measuring the FG frequency. Record its measured frequency and repeat for the other FG settings listed. For the 1 MHz signal, use the sweep magnifier to expand the trace horizontally. Observe the effect of the horizontal position control with the sweep magnifier on. Estimate the error in reading the period from the scope display and calculate the resulting error in the calculated frequency.

| FG freq. | Scope TB setting | div/cycle | error | calc. freq. | error |
|----------|------------------|-----------|-------|-------------|-------|
| 30 kHz | _____ | _____ | _____ | _____ | _____ |
| 750 kHz | _____ | _____ | _____ | _____ | _____ |
| 120 Hz | _____ | _____ | _____ | _____ | _____ |
| 1 MHz (no mag) | _____ | _____ | _____ | _____ | _____ |
| 1 MHz (mag on) | _____ | _____ | _____ | _____ | _____ |

Question 1. Consult the operator's manuals for the FG and scope and determine if the agreement between the FG setting and scope measurement of frequency is within the manufacturer's specifications. Discuss the relative importance of errors such as: oscilloscope sweep rate inaccuracy and non-linearity, parallax error in scope observation, and calibration of the FG frequency dial.

Connect the TTL or SYNC output of the FG to the scope Ch 2 input. If your scope has a mode control, select "dual trace" or "chop" in order to activate both channels. Set the scope Ch 2 input to DC and the FG frequency to 10 kHz. Adjust the sensitivity and position controls to observe the Ch 2 input. Describe or sketch the waveform observed and its phase relationship to the sine wave signal on Ch 1. Repeat for the square wave and triangular wave output.

TTL and sine-wave outputs:

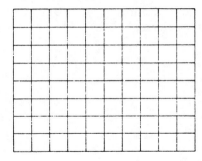

22

TTL and square—wave outputs:            TTL and triangular wave outputs:

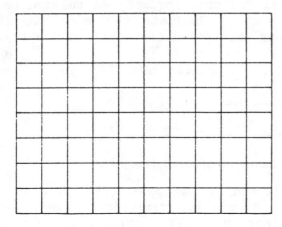            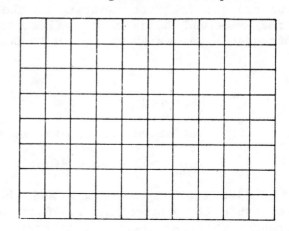

Graph 1.  On the basis of your measurements above, sketch the time relationship of all 4 FG outputs on a single graph.

Observe the dual trace display of signals in the frequency range of 1 to 10 kHz using sweep speeds of 2 ms/div to 200 µs/div.  If the selection of chopped and alternate modes is manual, what is the best sweep speed to change from one to the other?  If the mode change is made automatically, can you tell the sweep speed of changeover?  If not, consult the manual.

Fastest sweep speed for chopped mode is _____.

Source of information _____.

2-2  OSCILLOSCOPE TRIGGERING

Objectives:  To demonstrate the various trigger settings, modes, and sources which aid in the most effective applications of the scope by noting the effects of the trigger level and scope controls, by observing conditions of unstable triggering, and by using an external trigger signal.

Obtain a display of a 6 V peak-to-peak 1 kHz sine wave from the FG on Ch 1 and display the TTL output of the FG on Ch 2.  Use a sweep rate of 500 µs/div.  Set the sweep trigger controls to AC input, AUTO mode, Ch 1 source, positive slope.  Set the horizontal position control so that the start of the trace is visible on the screen.  Observe that the trace begins on the positive slope of the waveform.  What is the effect of varying the trigger level control?

Effect:

Change the trigger to negative slope and describe the effect on the waveform observed.

Effect:

Now select normal (rather than auto) triggering.  Set the Ch 1 input to GND. What happens to the display?  Is there any trigger signal with the Ch 1 input grounded?  Set the sweep trigger to AUTO and describe the traces.

Observations:

Adjust the FG frequency to obtain an almost stable trace.  Record this and several other frequencies below 1 kHz for which the trace is stable?  Include the lowest frequency for which the display is stable.

Stable display frequencies _____ Hz, _____ Hz, _____ Hz.

Question 2.  When there is no trigger signal, under what conditions is a stable trace obtained?  Why is the trace not completely stable?

Switch the Ch 1 input back to DC.  Set the sensitivity to 5 V/div.  Reduce the FG amplitude until the triggering fails in the AUTO mode.  Record the peak-to-peak deflection at this amplitude.

Deflection _____ div.

Keep the Ch 1 sensitivity at 5 V/div. and return the sine-wave amplitude to 6 V p-p.  Select the trigger source to be Ch 2, positive slope.  Obtain a stable trace and sketch the relative position of the first two cycles of each channel.  Move the FG TTL signal from Ch 2 input to the external trigger input.  Select external triggering source, positive slope, and adjust the level for stable display.  Compare the start of the Ch 1 trace with that observed before.

Reduce the FG amplitude to its lowest level.  Briefly explain why no loss of triggering occurs.

Explanation:

## 2-3  LOW-PASS AND HIGH-PASS FILTERS

Objectives:  To experience the transmission and attenuation characteristics of high-pass and low-pass filters and relate these to the theoretical response by wiring high-pass and low-pass filters to be connected between the FG and the scope and by observing the differences in amplitude and phase between the filter input and output signals.

Wire the circuit shown in figure 2-1.  The FG and scope should be connected to the circuit through the BNC or banana connectors of the frame. Be sure common connections are made to both the scope and FG.  Connection to common is automatically made when a BNC connector is used to connect to the instrument or when instrument common is connected to a black banana jack.

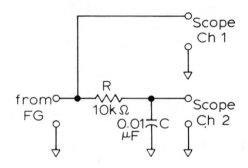

Fig. 2-1

Fill in the table of input and output amplitudes as a function of frequency below.  An easy way to do this is to set the scope time base so that dozens of cycles appear on the screen.  The signals then appear as horizontal bands whose height is a measure of the signal amplitude.  The FG output can be adjusted, if necessary, to keep the Ch 1 amplitude constant.  Use an FG amplitude of 5-10 V.  Express the ratio v(out)/v(in) in decibels (dB) (see text, page 34).

| Frequency | log frequency | v(in) | v(out) | v(out)/ v(in), dB |
|---|---|---|---|---|
| 10 Hz | _____ | _____ | _____ | _____ |
| 30 Hz | _____ | _____ | _____ | _____ |
| 100 Hz | _____ | _____ | _____ | _____ |
| 300 Hz | _____ | _____ | _____ | _____ |
| 1000 Hz | _____ | _____ | _____ | _____ |
| 3000 Hz | _____ | _____ | _____ | _____ |
| 10 kHz | _____ | _____ | _____ | _____ |
| 30 kHz | _____ | _____ | _____ | _____ |
| 100 kHz | _____ | _____ | _____ | _____ |

<u>Graph 2</u>.  Plot the transfer ratio in dB vs log frequency.  Determine the cut-off-frequency  and  the  attenuation  slope  (in  dB/decade)  at  high frequencies.

<u>Question 3</u>.  Compare the values of $X_C$ calculated from the nominal component values  with  the  value  of  $X_C$  calculated  from  the  experimental v(out)/v(in) ratio for each frequency above 300 Hz.  Explain any differences observed.

     With the FG at 100 kHz, set the scope time base to observe 1-5 cycles. Note that signal across the capacitor lags the signal across the function generator.  This phase shift can be measured as the fraction of a cycle delayed times $360^\circ$, i.e., (t(delay)/t(cycle)) x $360^\circ$.  What is the phase shift at 100 kHz?

$$\underline{\hspace{2cm}}^\circ \text{ at 100 kHz.}$$

Now measure the phase shift by means of Lissajous figures.  Set the FG to 10 Hz, and switch the scope display to X-Y to observe the Lissajous pattern as shown in figure 2-2.  Consult the manual for your scope to find how to select the X-Y mode.  Be sure to note the sensitivity setting of each input in your measurement.  The amplitude values should be recorded in volts rather than divisions.  Set the sensitivities so that the major axis of the ellipse is at an angle of about $45^\circ$ and several divisions in length.  The pattern should be centered on the screen so that the central chord of the ellipse c, can be measured with the vertical centerline of the scope graticule.

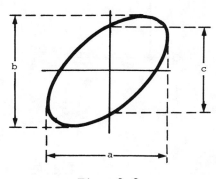

Fig. 2-2

An easy way to perform the measurement is as follows:

1)  Ground the vertical amplifier input (with the input switch) and align the trace with the horizontal center line.

2)  Switch the vertical amplifier to DC and ground the horizontal amplifier input.  Center the trace horizontally.  Measure the length of that trace which is the quantity b.

3)  Switch the horizontal amplifier to DC and measure c.  Repeat for the frequencies in the table below.

| frequency | log frequency | b,V | c,V | $\theta = \sin^{-1} c/b$ |
|-----------|---------------|-----|-----|--------------------------|
| 10 Hz | _____ | _____ | _____ | _____ |
| 30 Hz | _____ | _____ | _____ | _____ |
| 100 Hz | _____ | _____ | _____ | _____ |
| 300 Hz | _____ | _____ | _____ | _____ |
| 1000 Hz | _____ | _____ | _____ | _____ |
| 3000 Hz | _____ | _____ | _____ | _____ |
| 10 kHz | _____ | _____ | _____ | _____ |
| 30 kHz | _____ | _____ | _____ | _____ |
| 100 kHz | _____ | _____ | _____ | _____ |

Graph 3. Plot the phase angle vs. log frequency.

Question 4. Compare the measured phase angles with the theoretical value of $\theta = \tan^{-1} R/X_C$. Use $X_C$ values obtained from both nomimal component values and values obtained in question 3. Discuss your results.

Wire the high-pass filter of figure 2-3 by interchanging the R and C of the low-pass filter. Find the frequency of the input signal for which the output/input amplitude ratio is 0.707. (When the input amplitude on the scope is 4.25 divisions and the output amplitude is 3.0 divisions, the ratio is 0.707.) Record the cut-off frequencies for the high-pass filters given in the table below.

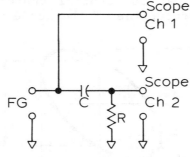

Fig. 2-3

| R | C | f1(nominal) | f1(observed) |
|---|---|-------------|--------------|
| 1 k$\Omega$ | 0.01 μF | _____ | _____ |
| 100 k$\Omega$ | 0.01 μF | _____ | _____ |
| 270 k$\Omega$ | 0.001 μF | _____ | _____ |
| 39 k$\Omega$ | 0.001 μF | _____ | _____ |

Question 5. Determine if the differences between the nominal and observed values are within the expected tolerance of the component values. Show your calculations.

Obtain an unknown filter from your instructor. Determine whether it is high-pass or low-pass and the value of its cut-off frequency. Determine the values of R and C in the filter. (Measure the value of R with an ohmmeter and calculate C from the value of RC obtained from the measurement of f(cutoff).)

Unknown No. _____

Filter type _____

f(cutoff) _____ Hz.

R _____ Ω

C _____ μF

## 2-4  FREQUENCY EFFECTS AT THE OSCILLOSCOPE INPUT

Objective:  To observe the effect of selecting the ac input on the scope by observing the high-pass filter behavior of the ac scope input and by determining the lower cut-off frequency of the ac input circuit.

When the AC input of an oscilloscope is selected, a capacitor is inserted to block the dc component of the input signal. The result is a high-pass RC filter at the scope input in which R is the scope input resistance.

With the FG connected to the scope, observe the difference in the effect of the FG offset control when the scope vertical input is in the AC and DC positions. This difference is most easily observed by connecting the FG output to both Ch 1 and Ch 2 and switching Ch 1 to DC input and Ch 2 to AC input. Closely observe the ac-coupled trace when a sudden change in offset is applied. Briefly describe the effects observed.

Effects:

Measure the cut-off frequency of the ac-coupled input by adjusting the FG frequency to the value for which the ratio of the ac to dc-coupled deflections is 0.707. (This frequency is probably very low.)

_____ Hz

Question 6.  Use the nominal value of the scope input resistance for R and calculate the value of the capacitor in the ac input of the scope.

## 2-5  SHIELDING AND INDUCED SIGNALS

Objective:    To develop an appreciation for the degree to which signals can be
              induced in wires connected to nothing other than the high-
              resistance input of a measuring device by the observation of
              induced signals in the connections to the scope through the
              breadboard frame.

    Disconnect all signal inputs and outputs from the breadboard frame.
Connect a shielded cable to the scope input, but leave the other end of the
cable unconnected.  Observe the display and measure the amplitude of any
signal present.  Connect the cable to the breadboard frame and again measure
the signal amplitude.  Now insert a job board and again measure the signal
amplitude.  Also determine the signal frequency.  The signal can be further
increased by connecting a wire to a job board contact to the scope connector
and touching the other end of the wire.  The amplitude of the signal is also
affected by whether you are touching the frame (or other ground) or not.

                    Open circuit signal amplitude _____ V.

Question 7.    What is the source of the signal observed and why is it affected
               by the amount of unshielded conductor exposed?

    Connect the FG signal output to a frame connector.  Connect the FG output
to the Ch 1 scope input through the contacts on a job board.  Set the FG for a
1 kHz, 10 V sine wave.  Connect the Ch 2 scope input to another frame
connector, but do not make any connections to the corresponding job board
contact.  Record the magnitude of the signal observed on channel 2 at the
frequencies in the table below.  Keep the Ch 1 amplitude constant at 10 V.
Now connect a 10 kΩ resistor between the Ch 2 input contact on the job board
and common.  Repeat for a resistor of 1 kΩ.

| frequency | Ch 2 | Ch 2 amplitude, 10 kΩ | Ch 2 amplitude, 1 kΩ |
|-----------|------|-----------------------|----------------------|
| 10 Hz | _____ | _____ | _____ |
| 100 Hz | _____ | _____ | _____ |
| 1 kHz | _____ | _____ | _____ |
| 10 kHz | _____ | _____ | _____ |
| 100 kHz | _____ | _____ | _____ |
| 1 MHz | _____ | _____ | _____ |

Comment:  What you have observed is called cross-talk.  It occurs when a
signal in one conductor can induce a signal by capacitive, inductive or
electromagnetic coupling in another conductor.  The small currents induced in
the "receiving" conductor produce voltages across the impedance between the
conductor and common.  The larger the impedance, the larger the voltage as
your data shows.  When high impedance signal lines must be used, they should
be shielded.  Examples of shielded lines used in high-impedance voltage
measurements are BNC cables and oscilloscope probes.  (See text, pp 40-41 for
use of scope probe.)

2-6  AC SCALES OF THE DIGITAL MULTIMETER

Objective:  To show that the response of the DMM to ac signals depends on the waveform measured and its frequency by simultaneously monitoring the FG output with the scope and DMM and comparing the amplitudes measured by each.

Connect both the scope and the DMM to the signal output of the FG. Compare the scope deflection (p-p) and the DMM "AC VOLTS" reading for the conditions given in the table below.

| frequency | waveform | amplitude, V p-p | FG offset, V | Calc RMS | DMM |
|-----------|----------|------------------|--------------|----------|-----|
| 100 Hz  | sine     | 1 V  | 0    | _____ | ____ |
| 100 Hz  | sine     | 5 V  | 0    | _____ | ____ |
| 100 Hz  | sine     | 10 V | 0    | _____ | ____ |
| 1 kHz   | sine     | 10 V | 0    | _____ | ____ |
| 10 kHz  | sine     | 10 V | 0    | _____ | ____ |
| 100 kHz | sine     | 10 V | 0    | _____ | ____ |
| 500 kHz | sine     | 10 V | 0    | _____ | ____ |
| 100 Hz  | sine     | 10 V | +5 V | _____ | ____ |
| 100 Hz  | sine     | 10 V | -5 V | _____ | ____ |
| 100 Hz  | square   | 10 V | 0    | _____ | ____ |
| 100 Hz  | triangle | 10 V | 0    | _____ | ____ |
| 100 Hz  | TTL      |      |      |           |      |

Question 8.  Show that the calculated RMS values for the sine wave signals with no offset agree or disagree with the DMM readings within the expected errors for the instruments.

Question 9.  On the basis of the effect of dc offset on the DMM reading, discuss the probable method of input coupling for the ac sales of the DMM.

Graph 4.  Plot the DMM response (as a fraction of the 100 Hz reading) vs. log frequency. Use the 10 V p-p sine wave with no offset. What is the cut-off frequency of the meter response?

Question 10.  Discuss the response of the DMM to the non-sinusodial waveforms. Use calculations of the expected readings in your discussion.

# UNIT 3.  POWER SUPPLIES

The diode is introduced in these experiments, and its application in practical power supplies is developed.  The current voltage curves of several diode types are observed.  Several rectifier circuits are wired and tested. The capacitive filter is developed as a voltage smoothing circuit and IC and Zener diode voltage regulation are studied.

|  Equipment:  |  Parts:  |
|---|---|

<table>
<tr><td>1.  Basic station</td><td>1.  Resistors:  10 Ω, 1 kΩ,<br>1.2 kΩ, 2 kΩ, 2.2 kΩ</td></tr>
<tr><td>2.  Utility job board</td><td>2.  Diodes:  signal, power,<br>Zener (≈ 5 V)</td></tr>
<tr><td>3.  Power supply job board</td><td>3.  Bridge rectifier package</td></tr>
</table>

## 3-1  DIODE CHARACTERISTICS

Objective:  To determine and compare the current-voltage curves for three types of semiconductor diodes (signal, power, and Zener) by connecting them into a simple curve tracer which uses the FG to continuously vary the current through the diode and uses the oscilloscope (X-Y mode) to display the current-voltage curve.

On the utility job board, connect the current voltage curve tracer circuit shown in figure 3-1.  Use the loose components provided.

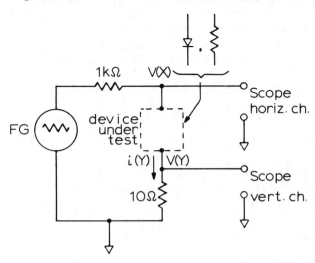

Fig. 3-1

31

The 1 kΩ resistor is used to limit the current in the circuit to a safe value for the function generator (FG). The 10Ω resistor functions as a current sensing resistor by providing a voltage proportional to the circuit current. The oscilloscope is operated in the X-Y mode so that the vertical deflection is proportional to the current through the device under test. The horizontal deflection is proportional to the voltage across the test device plus the voltage across the current-sensing resistor. A small resistor has been used for R(sense) to minimize its contribution to the horizontal deflection voltage. Thus the current-voltage curve of the test device is displayed on the oscilloscope as the FG output voltage varies between its peak positive and negative values.

Set the oscilloscope horizontal sensitivity at 2V/div and the vertical sensitivity at 10 mV/div. (For R(sense) = 10 Ω, this setting gives 1 division of vertical deflection per mA of current through the device). With both inputs switched to GND, center the dot in the scope display. Set the FG for a 100 Hz triangular signal. Connect a 2.2 kΩ resistor as the device under test, switch the scope horizontal input to dc, and adjust the FG amplitude and offset until the horizontal line displayed is 6 div. long and centered. Switch the vertical input to DC. Record the horizontal and vertical voltages for 2 points on the line. The voltage V(X) is V(X) = V(test) + i(Y) R(sense) and the current i(Y) is i(Y) = V(Y)/R(sense).

| horiz div | vert. div | V(X), V | i(Y), mA |
|-----------|-----------|---------|----------|
|           |           |         |          |
|           |           |         |          |

Question 1. Calculate the slope of the line, i.e., i(Y)/V(X). The slope of the line has the units A/V = $\Omega^{-1}$. At any point on the line V(X) = i(Y)(R(test) + R(sense)). From this equation, calculate the slope expected from the nominal value of the components. Compare with the measured value. What error is introduced if R(sense) is assumed to be negligibly small (zero)?

Replace the test resistor in the curve tracer with a signal diode. Connect the diode so that the banded end (the cathode) is connected to R(sense). The vertical deflection (voltage drop across R(sense)) is proportional to the current through the diode while the horizontal deflection is the voltage across the diode (plus a small error due to the voltage drop across R(sense)). Sketch the general shape of the curve. Label the axes in mA and V.

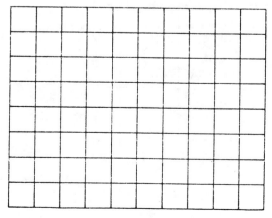

The exponential behavior of the diode can be observed by expanding the portion of the curve near zero. Change the horizontal sensitivity to 0.1 V/div and place the 0,0 point of the display at a graticule intersection in the lower left-corner. Adjust the FG amplitude and offset so that the diode characteristic curve fills the screen. If a second trace appears, reduce the FG amplitude and readjust the offset. Record voltages from the current/voltage curve for the current values in the table below.

| Current | Voltage |
|---------|---------|
| 0 mA | _____ |
| 0.5 mA | _____ |
| 1 mA | _____ |
| 2 mA | _____ |
| 3 mA | _____ |
| 4 mA | _____ |
| 5 mA | _____ |
| 6 mA | _____ |

Graph 1.  Plot the current-voltage curve for the diode. Plot also points corrected for the iR drop across R(sense). Use the corrected values to plot log i vs. V for the diode. Calculate the slope of the log i vs. V plot.

With the curve tracer at this same setting, compare the forward voltage drops of the three diode types below for a 5 mA forward current. (Diode numbers generally begin with 1N such as 1N914, 1N4735, etc.)

| diode type | diode # | V(X) | V(diode) |
|------------|---------|------|----------|
| signal | _____ | _____ | _____ |
| power | _____ | _____ | _____ |
| Zener | _____ | _____ | _____ |

Return the curve tracer to 2V/div horizontal deflection and return the 0,0 point to the center of the graticule. Use the test resistor to obtain a horizontal amplitude of 6V. Now sketch the characteristic curves obtained when the test device is the power diode and also when the Zener diode is the test device.

Power diode curve                Zener diode curve

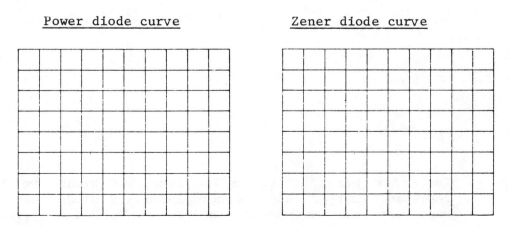

Disconnect the curve tracer. It will not be used again in this unit.

A diode can be checked with an ohmmeter. The polarity of the diode can also be determined when the polarity of the ohmmeter leads is known.

Connect the common lead of an ohmmeter to the cathode (banded) end of the power diode. Connect the other ohmmeter lead to the other terminal of the diode. Record the resistances measured on each resistance range of the ohmmeter. Reverse the leads and repeat.

| R range | $\Omega$, cathode common | $\Omega$, anode common |
|---------|--------------------------|------------------------|
| _____ | _____ | _____ |
| _____ | _____ | _____ |
| _____ | _____ | _____ |
| _____ | _____ | _____ |
| _____ | _____ | _____ |

<u>Question 2.</u>   Which lead from the ohmmeter is positive and how do you know this?

<u>Question 3.</u>   Why does the apparent resistance change with the resistance range selected?

Obtain an encapsulated rectifier bridge and use the ohmmeter to determine the internal arrangement of the four diodes which are connected between the four terminals of the device. Sketch the internal circuit below. Identify the terminals by the designations used on the package.

<u>Question 4.</u>   There are two contacts in the rectifier bridge which give a resistance larger than the normal forward diode resistance but much less than the reverse resistance. Which leads are they and why is the resistance higher?

## 3-2 POWER TRANSFORMER VOLTAGES

Objectives:  To become familiar with the power transformer connections through the breadboard frame to the power supply job board and to observe the power transformer secondary waveforms by connecting the power supply job board into the frame and observing the transformer secondary signals with the oscilloscope.

Obtain a power supply job board. When plugged into the right-most position of the breadboard frame, the power supply job board provides contact to the secondary of a 12.6 V power transformer. The arrangement of the connections is shown in figure 3-2. Do not remove any of the components from the power supply job board. With the power off, connect transformer contact C to common. Connect a scope input to transformer contact A. Be sure that no bare wire is exposed above the breadboard socket so that no shorts can occur between adjacent contacts. Be very careful to make the correct contact--the three contacts are all adjacent. Turn the power supply on and measure the peak-to-peak voltage and the frequency of the voltage at the transformer secondary. If it is not possible to get the entire waveform on the scope screen, use an attenuator scope probe or estimate the p-p voltage by measuring the peak voltage.

V(A-C) = _____ V, _____ Hz

Method used to measure V(p-p) _____

Turn the power supply off while changing the scope lead from contact A to B. Always turn the power off to make wiring changes on any job board. Measure the p-p voltage between contacts B and C.

V(B-C) = _____ V.

Move the common contact from C to B. (Did you remember to turn the power off?) Connect one scope channel to contact A and the other channel to contact C. Observe both signals and measure the p-p voltage of each. Also measure their relative phase.

V(A-B) _____

V(B-C) _____

Phase angle _____

Question 5.  Calculate the RMS voltages between each pair of transformer secondary contacts. How does this compare with the rated value?

# Power Supply Job Board

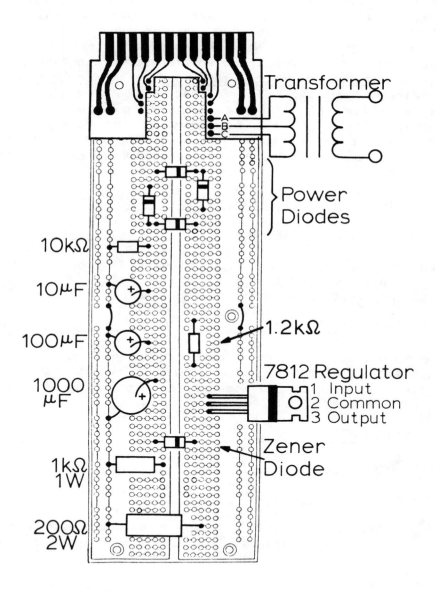

Fig. 3-2

## 3-3   RECTIFIER CIRCUITS

Objectives:   To study half-wave, full-wave, and bridge rectifier circuits and
their characteristics by connecting them on the power supply job
board and observing their input and output waveforms with the
oscilloscope.

Connect the half-wave rectifier circuit shown in figure 3-3 by connecting
patch wires to the appropriate components on the PS job board.  Use one of the
power diodes.  Remember that the PS job board must be plugged into the frame
at the far right to connect to the transformer secondary.

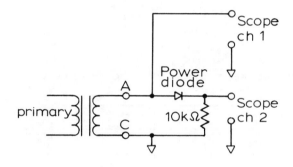

Fig. 3-3

Sketch the input and output waveforms for the half-wave rectifier.  Indicate
the signal amplitudes in volts.

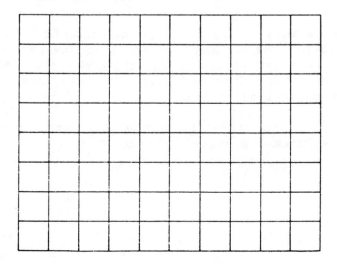

38

Disconnect the half-wave rectifier and connect the full-wave rectifier circuit shown in figure 3-4. Use power diodes. Note that two of the diodes on the PS job board already have their cathodes connected together.

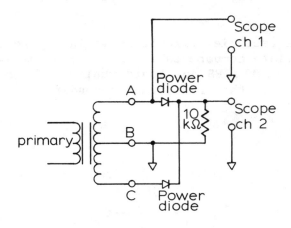

Fig. 3-4

Sketch the input and output waveforms for the full-wave rectifier and indicate the signal amplitudes.

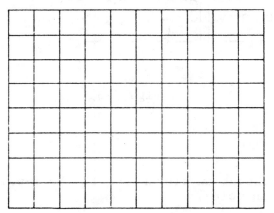

Question 6. The full-wave rectifier can be thought of as two half-wave rectifiers that conduct during alternate half cycles. Show from the phase measurements of the transformer secondary in experiment 3-2 that the diodes connected to transformer contacts A and C cannot conduct at the same time.

Question 7. Why is the voltage from the full-wave rectifier half that obtained from the half-wave rectifier?

Disconnect the full-wave rectifier and connect the bridge rectifier circuit of figure 3-5. (Note that the four power diodes on the PS job board are arranged in this configuration. You must jumper the anode to cathode connections, however). Use the 10 kΩ resistor provided on the PS job board.

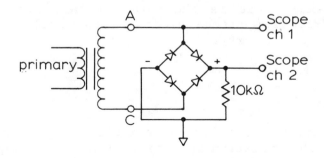

Fig. 3-5

Sketch the output waveform (the channel 2 display) and note the signal amplitude.

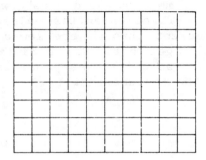

The signal at channel 1 is the voltage across one of the diodes in the bridge. Observe the complete waveform with careful attention to the position of zero so you can see that the waveform extends both above and slightly below zero volts. Measure the positive peak and negative peak voltages and sketch the waveform. It will be necessary to increase the vertical sensitivity to measure the negative peak voltage (relative to zero). Be sure to re-establish the scope zero point when changing sensitivity.

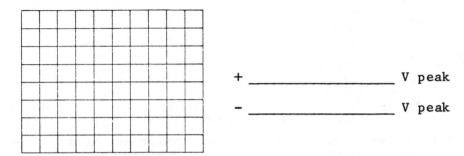

+ _____ V peak

- _____ V peak

<u>Leave the bridge rectifier connected for the next experiment.</u>

Question 8. During which part of the Ch 1 waveform is the diode conducting? Remember, you measured the voltage across, not the current through, the diode. What is the maximum forward voltage drop (the voltage drop during conduction)? What is the peak inverse voltage across the diode? Assuming all the diodes in the bridge are the same, what is the total voltage loss due to forward voltage drops across the diodes in the bridge rectifier? (See chapter 3 of the text).

## 3-4 CAPACITIVE FILTERS

Objectives: To investigate the characteristics of capacitive filters including the variation in output voltage and ripple voltage as a function of load by connecting various values of filters capacitors and load resistors to the bridge rectifier circuit.

Revise the bridge rectifier circuit to include the filter and load as shown in figure 3-6. Use components on the PS job board. Connect the -lead of electrolytic capacitors to common to avoid breakdown of the electrolyte and destruction of the capacitor. Three capacitor values and two load resistor values will be used. The current through the lower load resistor is quite large and this resistor will get hot. To minimize heating of the load resistor, make the measurements as quickly as possible and turn the power off immediately after completing the measurement.

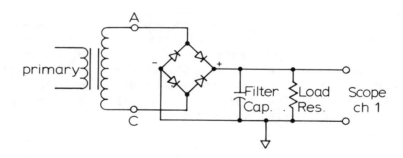

Fig. 3-6

Connect the first capacitor and load combination given in the table below. Measure V(out) with the Ch 1 scope input on DC. The measurement of the peak-to-peak ripple voltage, however, may require the use of AC coupling. The coupling capacitor blocks the dc component of the input signal so that a higher sensitivity can be used to measure the ac component. Measure V(out) and the ripple voltage for the other filter and load values in the table.

| filter capacitor | load resistor | V(out), peak | Ripple, p-p |
|---|---|---|---|
| 100 µF | 1 kΩ | _____ | _____ |
| 1000 µF | 1 kΩ | _____ | _____ |
| 10 µF | 1 kΩ | _____ | _____ |
| 10 µF | 200 Ω | _____ | _____ |
| 100 µF | 200 Ω | _____ | _____ |
| 1000 µF | 200 Ω | _____ | _____ |

__Question 9.__  Estimate the average dc output voltage as V(out) peak minus half the p-p ripple voltage. Compare the estimated average V(out) with the theoretical value of V(dc) = 1.4 V(rms)[1-1/(2fC(filter)R(load))] where V(rms) is the transformer voltage for each of the capacitor and load combination used. Note that f should be the frequency of capacitor charging which is not equal to the line frequency for a full-wave or bridge rectifier. Calculate the ripple factor for each of the filter and load combinations studied. Compare the experimental values with the theoretical value r = 1/[2√3 fC(filter)R(load)].

## 3-5  VOLTAGE REGULATION

Objectives:  To experiment with integrated circuit (IC) and Zener diode voltage regulators and to compare their behavior with the ideal voltage source by connecting these regulators to the bridge rectifier circuit and observing their effect on the output voltage stability and ripple voltage amplitude.

Connect the circuit of figure 3-7 using C = 1000 µF and R = 1 kΩ. Use the 7812, 12 V regulator on the PS job board. Measure the ripple on the output using Ch 2 of the scope.

V(ripple), p-p = _____ V

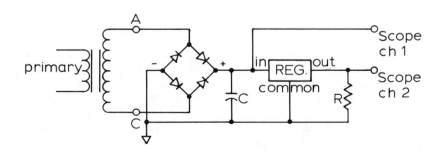

Fig. 3-7

Sketch both Ch 1 and Ch 2 outputs to show the relationship between the regulated and unregulated outputs. Use the DMM to measure the exact dc output voltage. Repeat the scope and DMM measurements for C = 100 µF, C = 10 µF and with C disconnected (no filter) for both 1 kΩ and 200 Ω loads.

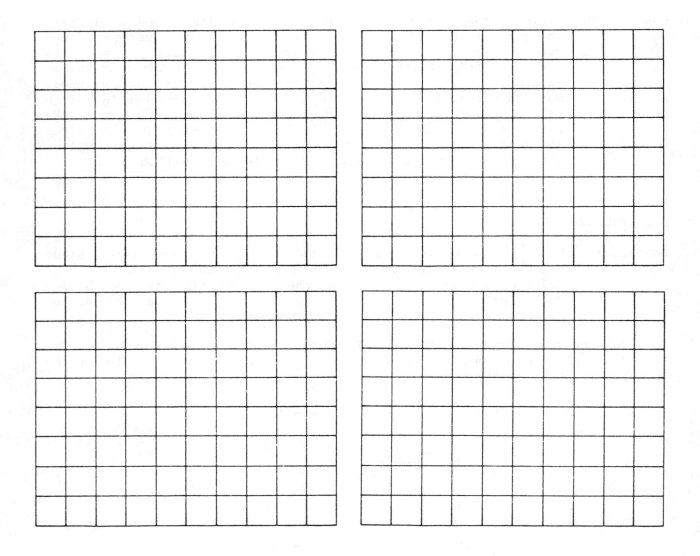

Question 10. Describe the conditions in the above measurements for which the regulator does not maintain a constant output. Suggest an explanation.

Connect the Zener diode shunt regulator circuit shown in figure 3-8.

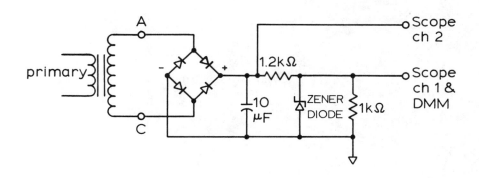

Fig. 3-8

Measure the output voltage with the DMM and the input and output ripple voltages with the scope.

V(out)(DMM) _____

Input voltage, max. _____

Input voltage, min. _____

Output ripple (p-p) _____

Question 11.  Calculate the current through the diode at the maximum input voltage [I(max)] and at the minimum input voltage [I(min)]. Calculate the effective resistance of the Zener diode for each voltage. The dynamic resistance of the Zener diode is equal to the Zener voltage variation divided by the corresponding change in diode current. Calculate the dynamic resistance of the Zener diode from the equation V(ripple)/[I(max)-I(min)]. This dynamic resistance is the effective output resistance of the Zener source. Comment on the Zener source as an ideal voltage source.

# UNIT 4. INPUT TRANSDUCERS AND MEASUREMENT SYSTEMS

The characterisitics of a variety of input transducers are studied in this unit. Included are the photovoltaic cell, the thermistor, the strain gauge and the phototransistor. The measurement process is not restricted to the magnitude of the transducer output (analog domain), but may involve the time relationships in the transducer output signal (time domains), or the use of the transducer for the detection of discrete events (digital domain). This unit thus introduces the "universal counter" or counter/timer/frequency meter (CTFM) for frequency and period measurements and events counting. Basic digital gates and flip-flops are also introduced in this unit as the basic building blocks of digital instrumentation and to extend your experience with operations on logic level signals.

<table>
<tr><td colspan="2"><u>Equipment</u></td><td colspan="2"><u>Parts</u></td></tr>
<tr><td>1.</td><td>Basic station</td><td>1.</td><td>Resistors: 100 Ω, 150 Ω,</td></tr>
<tr><td>2.</td><td>Counter/timer/frequency</td><td></td><td>220 Ω, 10 kΩ potentiometer</td></tr>
<tr><td></td><td>meter (CTFM)</td><td>2.</td><td>Transducers: photovoltaic cell,</td></tr>
<tr><td>3.</td><td>REF job board</td><td></td><td>thermistor, strain gauge</td></tr>
<tr><td>4.</td><td>BSI job board</td><td></td><td>assembly, opto-Interrupter,</td></tr>
<tr><td>5.</td><td>Basic gates job board</td><td></td><td>variable density filter,</td></tr>
<tr><td>6.</td><td>Flip-flop job board</td><td></td><td>cadmium sulfide photocell.</td></tr>
<tr><td>7.</td><td>Counter job board</td><td>3.</td><td>Miscellaneous: mercury</td></tr>
<tr><td>8.</td><td>Op amp job board</td><td></td><td>thermometer, styrofoam cup,</td></tr>
<tr><td>9.</td><td>Utility job board</td><td></td><td>water, ice</td></tr>
</table>

## 4-1 FREQUENCY AND PERIOD MEASUREMENT

Objective: To introduce the digital measurement of frequency and period by using the counter/timer/frequency meter (CTFM) to measure the frequency and period of the function generator output over a wide range of frequencies.

Connect the function generator (FG) TTL output to the A input of the counter-timer-frequency meter (CTFM) and to the scope. Set the CTFM to the frequency mode and the FG to 1 kHz. Measure the frequency both with the scope and with the CTFM. Consult the CTFM Instruction manual or operation guide for specific details of its operation. Set the time bases of the CTFM and the scope to get the highest measurement resolution. Now switch the CTFM to the period mode and measure the period of the same FG signal. Measure the frequency with the scope and CTFM of FG signals every decade from 1 Hz to 1

MHz. At each FG setting, measure the period of the signal with the CTFM on the highest resolution. Record your results in the table below.

Frequency

| Decade | Scope | Frequency meter | Period |
|--------|-------|-----------------|--------|
| 1 MHz | _____ | _____ | _____ |
| 100 kHz | _____ | _____ | _____ |
| 10 kHz | _____ | _____ | _____ |
| 1 kHz | _____ | _____ | _____ |
| 100 Hz | _____ | _____ | _____ |
| 10 Hz | _____ | _____ | _____ |
| 1 Hz | _____ | _____ | _____ |

What are the units of the number displayed by the CTFM in the frequency mode? _____. In the period mode? _____.

Disconnect the FG from the breadboard frame.

Question 1. For a 148 kHz signal, would a frequency or a period measurement with the CTFM provide the greater resolution? Explain.

Question 2. A particular model of CTFM has a 10 Hz time base in the frequency mode and 10.0 MHz time base in the period mode. At what frequency would the period and frequency measurements have equivalent precision for this instrument? Explain.

4-2 DIGITAL TIME BASE

Objective: To become familiar with the outputs and controls of the digital time base on the reference job board by observing the frequencies of the several outputs for each position of the time base control switches.

Install the reference job board in the frame. Refer to figure 4-1a for the location of the time base section of the reference job board and to figure 4-1b for the location of the control switches (C1-C4) and the digital time base outputs (1 MHz, ÷ N and ÷ NM).

Connect the CTFM and scope to one of the 1 MHz time base outputs. Confirm that the output frequency is 1.00 MHz. By using the scope in the dual trace mode, confirm that the two 1 MHz outputs are complementary (opposite in logic level). Measure the ÷N output frequency for the control switch positions indicated in the following table. Tabulate all of the ÷N output time base periods and values of N rounded to the nearest whole number.

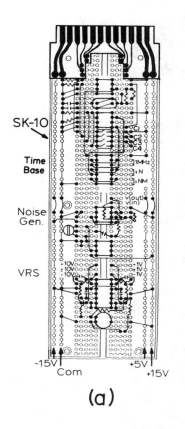

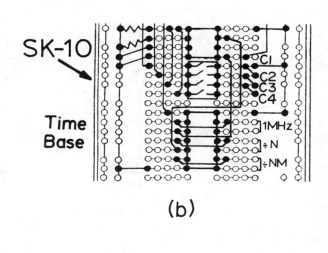

(b)

(a)

Fig. 4-1

| Control switch positions | | Measured frequency | Calculated period | Calculated N |
|---|---|---|---|---|
| C1 | C2 | | | |
| off | off | _____ | _____ | _____ |
| off | on | _____ | _____ | _____ |
| on | off | _____ | _____ | _____ |
| on | on | _____ | _____ | _____ |

As shown in figure 4-2, the time base outputs are derived from a 1 MHz crystal oscillator and two frequency divider IC's controlled by switches C1-C4.

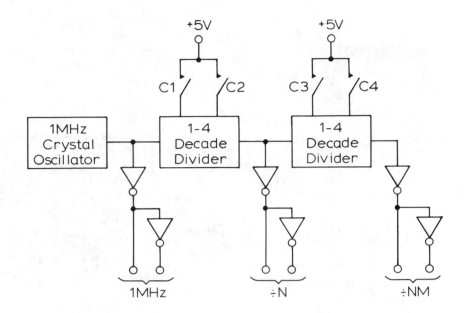

Fig. 4-2

Set control switches C1 and C2 so that N = 1,000. Connect the CTFM to the ÷NM output. Use the period mode to measure the output time base for the positions of switches C3 and C4 indicated below. Report the measured periods and the calculated values of M for each switch position.

| Control switch position | | Measured period | Calculated M |
|---|---|---|---|
| C3 | C4 | | |
| off | off | _____ | _____ |
| off | on | _____ | _____ |
| on | off | _____ | _____ |
| on | on | _____ | _____ |

Set switches C3 and C4 to give an M value of 10. By changing the position of switches C1 and C2, verify that the output is divided by N x M.

Question 3. For C1-C4 settings of off, off, off, off, what is the ÷NM time base output period and frequency?

Question 4. It is recommended that the time base switches be set to provide 1 MHz, 1 kHz, and 1 Hz simultaneous outputs. What positions of C1-C4 are required to achieve this?

## 4-3 BASIC DIGITAL DEVICES

Objectives: To introduce the TTL integrated circuits that perform the basic logic functions of AND, OR, and INVERT by observing their response (truth table) to all possible combinations of input levels, and to introduce a JK flip-flop integrated circuit by noting the effects of the clock, J, K, and clear inputs on the changes in the output level.

Install the basic gates job board in a frame position adjacent to the binary source and indicator (BSI) job board. Refer to figure 4-3 for the location of the 7408 AND gate, the 7432 OR gate and the 7404 INVERT gate integrated circuits (IC's) on the job board. Locate the AND gate at the bottom of the board. Note that there are four two-input AND gates on the chip as seen by the pin diagram (pinout) in figure 4-3.

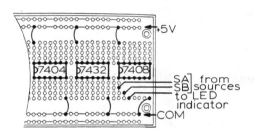

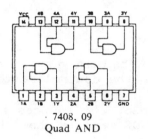

7408, 09
Quad AND

Fig. 4-3

The convention for identifying the pins on IC's is to number them counter clockwise from pin 1. Pin 1 is located by a notch or a small circular depression (or dot). The most common IC's are manufactured in 8 pin, 14 pin, and 16 pin packages. Large scale integrated (LSI) circuits are produced in packages of 18 to 40 pins.

Connect two logic level sources to one of the 7408 AND gate inputs and a LED indicator to the corresponding output as shown in figure 4-3. If desired, the switches can also be connected to LED indicators. Set the switches to the four possible combinations of logic states, observe the results, and fill in the table of states.

| Inputs | | Output |
|--------|--------|--------|
| SA | SB | M |
| L | L | ___ |
| L | H | ___ |
| H | L | ___ |
| H | H | ___ |

50

For the AND gate the output is LO whenever one or more inputs is
_____?

The AND gate will now be used to control the transmission of a waveform. Disconnect switch SB from the AND gate and connect the TTL output of the function generator to the input of an AND gate as shown in figure 4-4. Use switch SA as a gate control.

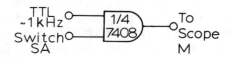

Fig. 4-4

Record your observations for both the switch positions.

Question 5. Explain what happens at the gate output for the two positions of the control switch.

Disconnect the sources and indicators from the AND circuit. The pinout of the 7432 OR gate is shown below. Verify its table of states.

| Inputs | | Output |
|--------|------|--------|
| SA | SB | M |
| L | L | _____ |
| L | H | _____ |
| H | L | _____ |
| H | H | _____ |

7432
Quad OR

For the OR gate the output is HI if one or more inputs is
_____?

The pinout of the 7404 inverter is shown below. Verify its table of states.

| Input | Output |
|-------|--------|
| SA | M |
| _____ | _____ |
| _____ | _____ |

7404, 05, 06, 16
Hex INVERT

Remove the basic gates job board before preceeding to the rest of the experiments. Obtain the flip-flop job board, and insert the board into the frame in place of the basic gates job board. Locate the dual 7473 J-K flip-flop on the job board by referring to the job board layouts in appendix B. Refer to figure 4-5, and connect logic source SA to a flip-flop clock (Ck) input. Also connect SA to one of the logic indicators. Connect the corresponding Q output of the flip-flop to a logic indicator as well. Toggle switch SA several times and observe the LED indicators. Make a table of the input and output states for at least 4 cycles of the input SA.

On which transition (LO → HI or HI → LO) of the CK input signal does the output change states? _____.

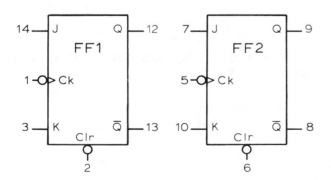 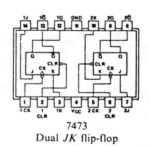

7473
Dual *JK* flip-flop

Fig. 4-5

In place of the switch connect the digital time base output of the REF job board to the Ck input and to an indicator. Set the time base for a 1.0 s period, and observe the relationship between the output and input waveforms.

What is the ratio of the input frequency to the output frequency? _____.

Leave the time base connected to the Ck input and a LED indicator connected to the Q output. Connect logic source SA to the J input of the flip-flop. Apply a LO to J while Q is LO and note the effect on the Q output. Apply a LO to J while Q is HI and note the effect below.

Effect:

Question 6. Explain the effect of a LO at the J input.

Disconnect SA from the J input and apply it to the K input. Repeat the previous experiment and note the result below.

Question 7. Explain the effect of a LO at the K input.

When nothing is connected to the J or K inputs, do the inputs behave as though they are LO or HI? _____.

Remove any connections to J and K, and connect switch output $\overline{MA}$ to the clear (CLR) input of the flip-flop. Apply a momentary LO to CLR when Q is in both states and note the effect.

Question 8. When does the clear signal take effect? Explain how a LO applied at CLR differs from a LO applied at J or K.

Remove the flip-flop job board before proceeding.

4-4 DECADE COUNTER, LATCH, AND DISPLAY

Objectives: To illustrate the functions of counting, latching, and decimal display and the use of the BCD code in decimal counting systems by connecting a seven-segment display through a BCD-to-seven-segment decoder to a source of BCD code and observing its response, then by using the decimal display to indicate the output of an IC decade counter.

Install the counter, latch, and display job board in the frame. Figure 4-6 shows the physical position and pin configuration of the display and decoder IC's on the counter job board.

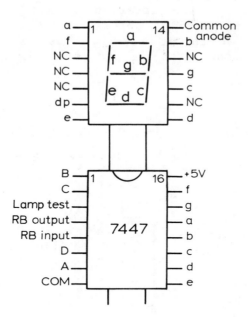

Fig. 4-6

Note carefully the display IC.  To light any given segment, the corresponding input should be connected to common through a current limiting resistor.  For example, to light number 1, segments b and c should each be connected to common through 100 Ω resistors.  The 7447 decoder/driver contains a switch to common for each of the seven segments.  The appropriate switches are closed in response to the BCD encoded number appearing at the A, B, C, and D inputs.  The logic by which the BCD numbers are converted to corresponding segment combinations will be studied in Unit 10.

Figure 4-7 shows the decoder and display connections for decimal display of BCD encoded information.   Note that this  diagram is a functional circuit

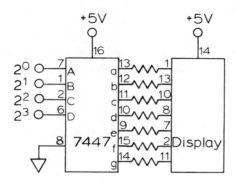

Fig. 4-7

diagram, where IC pin placement is made for convenience and bears no relationships to the actual pin location on the IC. Thus in a circuit diagram, pin 1 does not necessarily appear in the upper left hand corner of the IC block as it does on the IC itself. Instead pin 1 is placed in the circuit diagram wherever convenient for understanding the circuit operation.

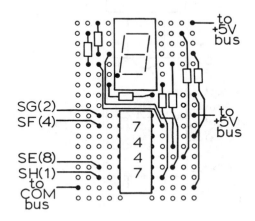

Fig. 4-8

Figure 4-8 shows how the wiring diagram of figure 4-7 has been translated into breadboard patch wires on the counter job board. Patch wire the switch outputs SE-SH to the decoder/driver IC as shown in the pictorial of figure 4-8. Note how this corresponds to the circuit diagram of figure 4-7 by identifying to which segment of the display the resistors on the pictorial figure are connected.

Mark on the pictorial diagram (figure 4-8), directly on the resistors, the display segment (a-g) to which each resistor is connected. Note that pins 4 and 5 of the display IC package are not connected (NC) to the IC inside. The breadboard contacts next to these pins are used as convenient multiple connectors.

Check your patch wiring for possible errors. Turn on the power to the job board. Set the switches to LLLL and note that the display shows the numeral zero. Note which segments (a-f) are lighted. Change the switch positions through all of the BCD codes, and check that the display shows the correct numerals. Identify in the following table which segments are lighted for each numeral.

| Numeral | BCD code | Segments |
|---------|----------|----------|
| 0 | LLLL | a,b,c,d,e,f |
| 1 | _____ | _____ |
| 2 | _____ | _____ |
| 3 | _____ | _____ |
| 4 | _____ | _____ |
| 5 | _____ | _____ |
| 6 | _____ | _____ |
| 7 | _____ | _____ |
| 8 | _____ | _____ |
| 9 | _____ | _____ |

**Remove** the **switch inputs** from the 7447 decoder.

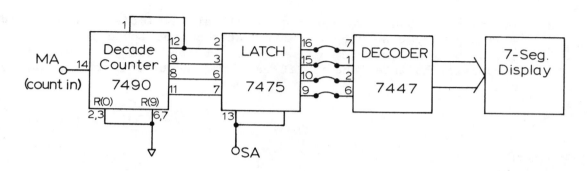

Fig. 4-9

Next the 7490 decade counter and 7475 quad latch will be connected to the decoder-display as illustrated in figure 4-9.
Connect the jumpers shown in the figure. Color coding the wire will make this easier. For example, use all blue wires for the four counter-latch connections, white for the latch-decoder connections, etc. Connect MA to the count input of the 7490. Connect the latch enable input to logic source SA, set SA to HI and apply power to the frame. Note the action of the counter when MA is toggled. Stop the counter when the display shows the numeral 1. Change the state of SA to LO, and depress MA two or three times. Change SA back to HI and note the result.

On what transition (HI → LO or LO → HI) does the counter output change? _____

Question 9. Describe the function of the latch enable signal.

Connect the function generator TTL output to the count input of the 7490. Use a 10 kHz input frequency and measure the frequency at QA (pin 12) and QD (pin 11) with the scope.

Input frequency _____

Frequency at QA _____

Frequency at QD _____

<u>Question 10</u>.   What are the divider fractions for the two outputs?

<u>Question 11</u>.   A counter has an upper frequency limit of 100 kHz and a signal
of 750 kHz is to be measured. How could the properties of the
7490 be used to advantage in this measurement?

Remove the counter job board from the frame.

4-5   SIGNAL LEVEL DISCRIMINATION

Objectives:   To introduce the comparator as a signal level discriminator and
waveshaper by observing the comparator output logic level as the
signal and threshold values are varied and by using the
comparator to interface the FG waveform output with the frequency
meter.

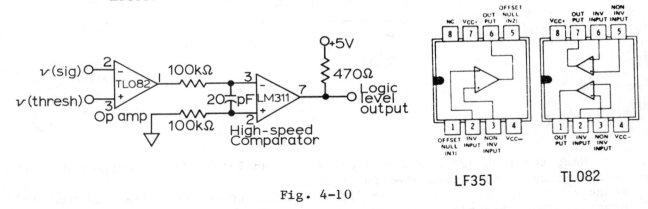

Fig. 4-10

The discriminator circuit used in this experiment is shown in figure
4-10.   The high-speed comparator IC in this circuit will accept an input
signal over a wide range ($\pm$ 15 V) and produce a logic level output (0.2 V and
+5 V).   For many applications, it is the only signal conditioning element
required.   However, its very fast response makes it susceptible to oscillation
when the input signal passes slowly through the threshold region.   The
operational amplifier (op amp), when used as a comparator, is slower and less
subject to oscillation, but it produces a $\pm$ 12 V output that is not compatible
with logic circuits.   The combination op amp and comparator shown in figure
4-10 provides stable logic level output transitions even for slowly changing
input levels.   The op amp and comparator IC's are found on the op amp job
board.

Install the operational amplifier (op amp) job board in the frame, and
locate the TL082 op amp and the LM 311 comparator shown in figure 4-11. Wire
the discriminator circuit as shown in figure 4-10.   The comparator input
resistors and capacitor and the output pullup resistor are prewired on the job
board.

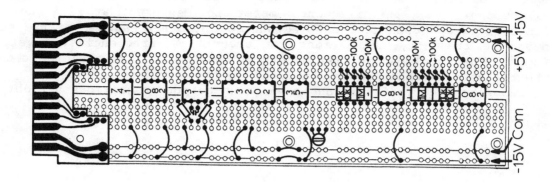

Fig. 4-11

Connect the $\pm$ 1 V output of the VRS to the v(sig) input of the discriminator. Connect the threshold input to common and the comparator output to an LED indicator on the BSI job board. Use the DMM to monitor the VRS voltage. Carefully adjust the voltage until the indicator just turns on. Similarly adjust the voltage until the LED just turns off. Note the voltages below.

       v(sig) for LED on  = _____

       v(sig) for LED off = _____

    Disconnect the LED indicator and the connections to both discriminator inputs. Connect the function generator to the signal input and to Ch 1 of the scope. Connect the $\pm$ 10 V VRS output to the threshold input of the discriminator. Connect the comparator output to Ch 2 of the scope. Set the function generator to produce a 1 kHz 10 V p-p triangular wave. Vary the threshold adjustment over the full $\pm$ 10 V range and note the effect. Sketch the waveforms for several settings of the threshold.

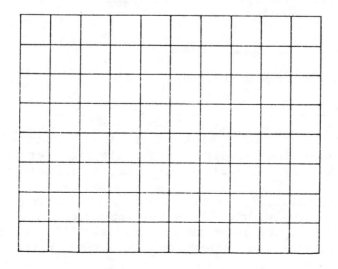

Question 12. Explain the relationship between the pairs of waveforms.

Remove the VRS from the threshold input, and connect the threshold to common. Connect the comparator output to the frequency meter and the oscilloscope. Use the frequency meter to measure the frequency of the triangular wave. Connect the FG output to the other channel of the scope and repeat the frequency measurement for several FG frequencies, waveforms and offsets. The oscilloscope display can be used to confirm the conditions under which reliable waveshaping is obtained.

Describe the conditions under which no output transitions are obtained.

Leave the op amp job board installed and the discriminator circuit connected.

Question 13. How could the discriminator threshold input be used to produce output transitions for a sine wave signal that varies between +1 V and +5 V peak-to-peak?

4-6  AN ENERGY CONVERSION TRANSDUCER

Objectives: To investigate the response of a photovoltaic cell to changes in light intensity and to develop a light detector by measuring the photovoltaic cell output and combining the photovoltaic cell with a discriminator circuit.

Obtain a photovoltaic cell and mount it on the utility job board. Set the DMM to the voltage mode and connect it to the cell. Cover the photovoltaic cell with an opaque object like your hand and measure the resulting signal.

V(dark) = _____

Now uncover the cell and record the cell voltage.

V(light) = _____

Question 14. Describe an experiment that would show the mathematical relationship (linear, logarithmic, etc.) between the photovoltaic cell voltage and the light intensity.

Connect the photovoltaic cell between the system common and the discriminator input. Connect the threshold input to the $\pm$ 1 V output of the VRS. Connect the comparator output to an indicator LED. Adjust the threshold level with the VRS so that the transducer/discriminator indicates whether the photocell is illuminated or in the dark.

Is the LED on or off when the photocell is illuminated? _____

**Question 15.** How would you change the circuit connections to reverse the state of the LED indicator?

**Question 16.** Explain how the photovoltaic cell/discriminator combination could be used to detect the passage of people through a door such as the entrance to an elevator or office.

## 4-7  RESISTIVE TRANSDUCERS

Objectives:  To observe the response of a thermistor to changes in temperature by measuring the thermistor resistance with a DMM and to obtain the response of a strain gauge to mechanical deformation by measuring its resistance change with a Wheatstone bridge.

Connect a thermistor to the DMM and set the function switch to the resistance mode. Allow a few minutes for the thermistor to equilibrate in room air and record the thermistor resistance Rt. Also measure the temperature with a mercury thermometer. Immerse the thermistor and the thermometer in a container of warm water (styrofoam cup if possible) and record the temperature and resistance. Add a small chunk of ice, stir until the ice is melted and record the results. Repeat for several additions of ice.

| T, $^{o}$C | Rt, $\Omega$ |
| --- | --- |
| ____ | ____ |
| ____ | ____ |
| ____ | ____ |
| ____ | ____ |
| ____ | ____ |
| ____ | ____ |

**Graph 1.** Plot the thermistor resistance versus temperature. Is there an obvious mathematical relationship between the variables? You might try $\log_2 Rt$ versus T, log Rt vs. 1/T (absolute temperature), and $(\log Rt)^2$ vs. 1/T.

Now connect the Wheatstone bridge circuit of figure 4-12(a). The 10 k$\Omega$ potentiometer serves as both Ra and Rb with point A being the wiper. The strain gauge assembly (Rsg) consists of a piece of tempered aluminum with a strain gauge (Vishay Measurements Group, type EA-13-250BB-120) attached with "super glue" as shown in figure 4-12(b). Apply power to the circuit and adjust the potentiometer until the DMM reads as close to 0.000 V as possible. Begin the adjustment on an insensitive DMM scale, and as null is attained select the most sensitive scale.

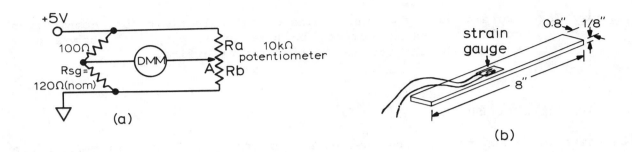

Fig. 4-12

Support the strain gauge assembly on the edge of the desk and push gently with your hand as illustrated in the figure.  Obtain the resistance for several different strains.

| Strain | Rsg |
|--------|-----|
| light  | _____ |
| medium | _____ |
| heavy  | _____ |

Question 17.  Strain gauges are widely used in industrial applications. suggest a possible use.

You may also wish to investigate a cadmium sulfide photocell for light intensity measurements.


4-8  LIMITING CURRENT TRANSDUCERS

Objectives:  To investigate a phototransistor and an opto-interrupter by observing their responses to changes in light intensity.

Connect the circuit shown in figure 4-13 on the utility job board.

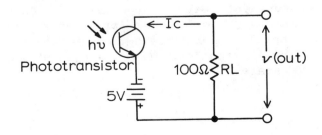

Fig. 4-13

Measure the output voltage v(out) for three different intensities of light striking the phototransistor. Calculate the collector current Ic for each light level.

| Intensity | v(out) | Ic |
|-----------|--------|-----|
| Bright | _____ | _____ |
| Medium | _____ | _____ |
| Dim | _____ | _____ |

Question 18. Calculate the change in the phototransistor bias voltage V(CE) in going from dim light to bright light. Why should the IR drop across the load resistor be small compared to the supply voltage?

Mount an opto-interrupter on the utility job board, and wire the circuit shown in figure 4-14. Connect the output of the opto-interrupter to an indicator light and to the DMM. Place an opaque card between the source and detector and measure the output voltage for the light and dark states.

V(light) = _____        V(dark) = _____

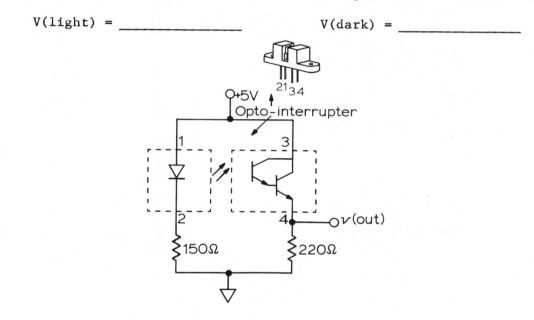

Fig. 4-14

If available, place a variable density filter between the source and detector and observe the change in output as the different regions of the filter are placed in the light path.

Question 19. Suggest a method for determining the linear velocity of a moving object using an opto-interrupter.

Question 20. Is the opto-interrupter a yes/no device, a proportional device or both? Explain.

# UNIT 5. OP AMPS AND SERVO SYSTEMS

The operational amplifier (op amp) is the key element in analog electronics circuits for measurement, control, signal processing and computation. This experiment introduces operational amplifier circuits for impedance buffering, signal amplification, current-to-voltage conversion, summing, integration and differentiation. Several important characteristics of op amps are explored and related to errors in op amp applications.

| Equipment | Parts |
|---|---|
| 1. Basic station | 1. Resistors: 47 Ω, 100 Ω |
| 2. CTFM | 2. Capacitors, 1 μF $\pm$ 10%, 1 μF $\pm$ 20%, |
| 3. Reference job board | unknown capacitor |
| 4. BSI job board | 3. Signal diode |
| 5. Op amp job board | |
| 6. Utility job board | |

## 5-1 NULL VOLTAGE MEASUREMENT

Objectives: To make comparison measurements of an unknown voltage and to become acquainted with the input/output characteristics of op amps and comparators by observing the output state as the magnitude and rate of change of the difference input voltage are varied.

Obtain the op amp job board shown in figure 5-1. The op amp job board contains one high quality IC op amp (LF351) with offset adjust, three medium quality IC op amps (TL082), one general purpose IC op amp (μA741), one quad IC analog switch (LF13202), one IC high speed comparator (LF311), and two precision resistor arrays. One array has decade values of resistors (1 kΩ to 10 MΩ), while the other array has one 10 MΩ, one 1 MΩ, two 100 kΩ, and two 10 kΩ resistors. The resistors in each array have one end connected in common for convenient use as op amp input and feedback elements.

Wire the circuit of figure 5-2 using one of the TL082 op amps. Note that the power connections to each IC on the op amp job board are prewired. Care should be taken to ensure that ICs are powered whenever an external signal is applied. Failure to do this will destroy most ICs. For v(unk) use the $\pm$10 V output of the VRS set at any value between +0.95 V and -0.95 V. Connect the $\pm$1 V VRS output to the inverting input of the op amp.

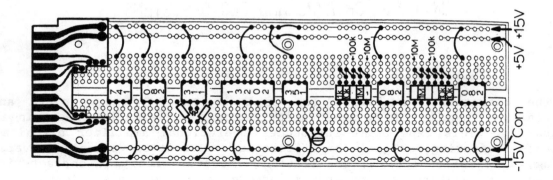

Fig. 5-1

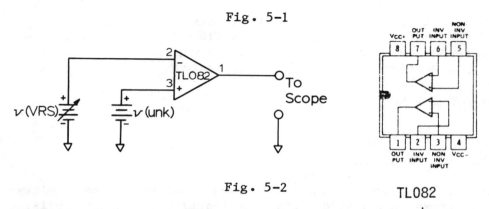

Fig. 5-2

TL082

Use the scope to monitor the op amp output and adjust the $\pm$1 V output of the VRS until the transition from one output voltage limit to the other occurs. Determine the positive and negative voltage limits of the op amp output.

+ limit = _____

- limit = _____

Leave the op amp comparison circuit just as it is and connect the DMM to the $\pm$1 V VRS output. Use the scope as an indicator to adjust the VRS (carefully) to the value where the op amp output just begins to decrease from its positive limit, where it is as close to zero output as you can set it and where it is not quite to the negative limit. Record these three values below. Repeat these observations several times.

| VRS value for: | Trial 1 | Trial 2 | Trial 3 |
|---|---|---|---|
| near + limit | _____ | _____ | _____ |
| near 0 | _____ | _____ | _____ |
| near - limit | _____ | _____ | _____ |

Now, without changing any settings, use the DMM to measure v(unk), the $\pm$10 V output of the VRS.

v(unk) = _____

Question 1. Estimate the op amp open loop voltage gain and the op amp input offset voltage from the above measurements.

Disconnect the circuit from the op amp and connect the similar circuit shown in figure 5-3 to the LM311 comparator IC. Use the 100 kΩ and 470 Ω resistors and the 20 pF capacitor that are already connected to the comparator IC on the job board. These components should remain in place. The LED indicator is one of the logic level indicators on the BSI job board.

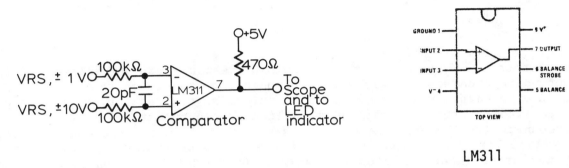

Fig. 5-3

Set the $\pm$10 V VRS output voltage to a value in the $\pm$1 V range and measure its value with the DMM. Carefully adjust the $\pm$1 V VRS output to determine the minimum change in input voltage required to complete the output transition from one state to the other. Approach the transition region first from one direction then the other. Record your data below.

v(unk), $\pm$10 V output  = _____

+ output voltage limit = _____

– output voltage limit = _____

|  | Trial 1 | Trial 2 | Trial 3 | Trial 4 |
|---|---|---|---|---|
| VRS, beginning of transition: | _____ | _____ | _____ | _____ |
| VRS, completion of transition: | _____ | _____ | _____ | _____ |

Question 2. Determine the window of uncertainty in the comparator input voltage and the input voltage offset of the comparator from the above measurements. Relate these to the precision and accuracy that would be obtained when this comparator IC is used for a null voltage measurement.

Compare the oscilloscope and LED output level indicators as the VRS voltage is adjusted slowly through the transition region. The oscilloscope will probably indicate some rapid oscillation in output states at some point in the transition region. Describe this behavior and tell what state the LED indicator is in during the oscillation.

Description of oscillation, including approximate frequency:

The output oscillation in the comparator response can be eliminated if the input voltage passes quickly enough through the transition region. The FG triangular wave output will now be used to determine the rate of input voltage change required to produce a single transition at the output.

Disconnect the VRS outputs from the comparator inputs. Instead, connect the FG triangular wave output to one of the comparator input resistors and connect the other input resistor to common. Be certain the comparator IC is powered before applying the FG signal. Adjust the FG to produce a 100 kHz, $\pm 1$ V peak-to-peak, triangular wave output. Connect the comparator output to the scope and the CTFM. Connect the FG TTL output to the scope external trigger input and set the scope time-base to trigger on this signal. Observe the output transition with the oscilloscope and measure the frequency with the frequency meter.

Decrease the FG output frequency until the frequency meter indicates an erratic and/or higher-than-expected frequency. Inspect the oscilloscope trace for indications of oscillation at the transition region. Record the frequency at which the transition first becomes unstable.

Highest frequency for unstable transition _____ Hz.
($\pm 1$ V triangular wave)

Now reset the FG to produce a $\pm 10$ V triangular wave output and repeat the above measurements.

Highest frequency for unstable transition _____ Hz.
($\pm 10$ V triangular wave)

Question 3. What is the slowest rate of change of the input voltage (in volts per second) that will produce a single output transition with your comparator IC?

## 5-2   VOLTAGE FOLLOWER

Objectives:   To wire, use, and characterize a voltage follower amplifier by
observing its response to input signal changes, by determining
its accuracy, and by measuring its output limitations.

Connect one of the TL082 op amps as a voltage follower as shown in figure
5-4.

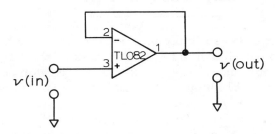

Fig. 5-4

Use the VRS for v(in) and the DMM to measure v(out) and v(in) for five or more
VRS settings in the range +10 to -10 V.   Calculate the gain from gain =
v(out)/v(in).

| v(in) | v(out) | Gain |
|-------|--------|------|
| _____ | _____ | _____ |
| _____ | _____ | _____ |
| _____ | _____ | _____ |
| _____ | _____ | _____ |
| _____ | _____ | _____ |
| _____ | _____ | _____ |

A 1 kΩ resistor will be used in series with the $\pm$10 V VRS output to
simulate a voltage source with a 1 kΩ internal resistance.   A 10 kΩ resistor
will be used to simulate the input resistance of the readout device used to
measure the source voltage.   Wire the circuit of figure 5-5a using the
resistors on the op amp job board, set the VRS output to a value between 1 V
and 2 V, and measure the voltage at point A with and without the 10 kΩ load
connected.

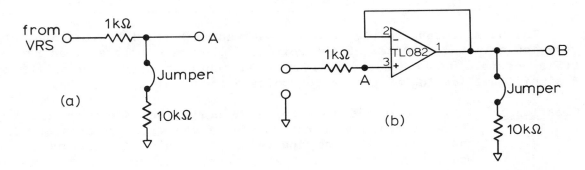

Fig. 5-5

        v(without load) = _____,
        v(with load)   = _____.

Now connect an op amp voltage follower (TL082) to buffer the voltage source from the load resistance as in figure 5-5b. Measure the follower output voltage with and without the 10 kΩ load resistor connected.

        v(without load) = _____,
        v(with load)   = _____.

<u>Question 4.</u>   A transducer has an output resistance of 80 kΩ. Describe how a follower could be used to decrease the loading error if the transducer output voltage is to be measured with a 1 MΩ input resistance oscilloscope and indicate the percent error avoided.

Disconnect the VRS from the follower and connect the follower input to the +10 VRS output. Connect a 1 kΩ load resistor R(load) to the follower output as shown in figure 5-6 and measure the output voltage with the DMM.

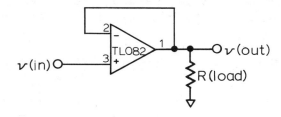

Fig. 5-6

Remove R(load) and measure the no load voltage. Repeat using the 100 Ω and 47 Ω load resistors. (Use loose parts for these values of R.) Calculate the output current.

| R(load) | v(out) with load | v(out) without load | i(out) with load |
|---------|------------------|---------------------|------------------|
| 1 kΩ | _____ | _____ | _____ |
| 100 Ω | _____ | _____ | _____ |
| 47 Ω | _____ | _____ | _____ |

The small output voltage change observed with the 1 kΩ load indicates the very low output resistance of the voltage follower. At lower load resistance, it is possible to exceed the maximum output current capability of the op amp. A significant loading of the output voltage occurs in such a case.

<u>Question 5.</u>   On the basis of the change in output voltage with a 1 kΩ load, estimate the output resistance of the voltage follower, or if no change was observed, calculate the value the output resistance would have if the smallest observable change had been measured.

Question 6.    Calculate the maximum output current the op amp can supply, based on your observations with the 100 Ω and 47 Ω loads.

Question 7.    Given that the TL082 BIFET op amp has an input resistance of $10^{12}$ Ω, explain how the voltage follower operates as a impedance buffer.  Calculate the ratio of the input to output resistance.

## 5-3  FOLLOWER WITH GAIN

Objectives:    To connect and characterize a follower-with-gain amplifier by comparing its observed and expected gains and noting the effects of amplifier input offset voltage.

Use the TL082 BIFET op amp to wire the follower with gain as illustrated in figure 5-7.  A convenient way to connect resistors to op amps is with the precision resistor arrays on the op amp job board.  Connect a 100 kΩ and a 10 kΩ resistor from the resistor array to the op amp as shown in figure 5-8.

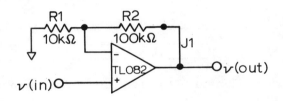

Fig. 5-7

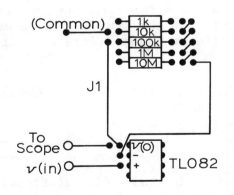

NOTE:  DO NOT remove the short jumper wires on the right side of the resistor array, or power jumpers to any components on this board.

(power jumpers are not shown)

Fig. 5-8

Use the VRS as v(in) and adjust the input voltage to approximately 0.2 V. Measure v(out), and calculate the gain.  Repeat for the other nominal v(in) values shown in the table.  Accurately measure both v(in) and v(out).

| v(in), nominal | v(in), measured | v(out) | Gain |
|----------------|-----------------|--------|------|
| 0.2 V | _____ | _____ | _____ |
| 0.4 V | _____ | _____ | _____ |
| 0.8 V | _____ | _____ | _____ |
| 1.0 V | _____ | _____ | _____ |
| 2.0 V | _____ | _____ | _____ |
| 5.0 V | _____ | _____ | _____ |

Question 8. Compare the measured gain to that expected from the nominal resistor values. Is the difference within the tolerance of the resistors?

Question 9. What limits the gain for large values of v(in)? Explain.

Adjust the $\pm 1$ V VRS to a voltage of slightly less than 10 mV. Now R2 will be changed to provide higher gains by connecting the movable contact J1 shown in figure 5-8 first to the 1 M$\Omega$ and then to the 10 M$\Omega$ resistor.

For each value of R2 measure v(out) and v(in) and calculate the gain.

| R2 | v(in) | v(out) | Gain |
|------|-------|--------|------|
| 100 k$\Omega$ | _____ | _____ | _____ |
| 1 M$\Omega$ | _____ | _____ | _____ |
| 10 M$\Omega$ | _____ | _____ | _____ |

Question 10. Compare the measured gains to those expected and explain any deviations.

Question 11. Can you obtain a gain of less than one with the follower with gain circuit? Explain why or why not.

At high gains an error may be observed due to the amplifier input offset voltage, v(offset). This effect can be illustrated by connecting the follower input to common and measuring v(out) for the same R2 values as above. Be sure to disconnect the VRS from the follower input before connecting the input to common.

| R2 | Gain, calc. | v(out) | v(offset) |
|------|-------------|--------|-----------|
| 100 k$\Omega$ | _____ | _____ | _____ |
| 1 M$\Omega$ | _____ | _____ | _____ |
| 10 M$\Omega$ | _____ | _____ | _____ |

Question 12. What percentage error in gain did v(offset) cause for the nominal gain of 100 amplifier of the previous experiment? How does the error due to v(offset) compare to the error resulting from resistor inaccuracy?

## 5-4  CURRENT-TO-VOLTAGE CONVERTER

Objectives:  To introduce the current follower amplifier, to characterize its
behavior and apply it in the measurement of small currents by
observing its response to known input currents, comparing the
observed and expected gain, and measuring the reverse bias
current of a signal diode.

Connect the TL082 as a current-to-voltage converter (current follower) as
shown in figure 5-9a.

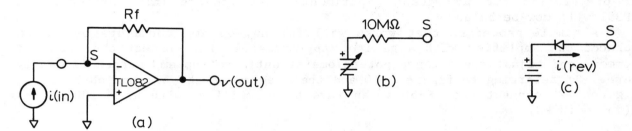

Fig. 5-9

Use a feedback resistor Rf of 1.0 k$\Omega$.  Use the 10 M$\Omega$ precision resistor in the
resistor array and wire the current source shown in figure 5-9b.  The 10 M$\Omega$
resistor in series with the VRS provides an input current i(in) equal to
$v(VRS)/10^7$.  Adjust the VRS to provide several input currents in the $10^{-7}$ to
$10^{-6}$ A range.  Measure the output voltage v(out) with the DMM.  Change Rf to 1
M$\Omega$ and later to 10 M$\Omega$.  Vary i(in) from nanoamperes to microamperes and
compare the measured output voltage to the expected values.  Do this for five
or more input currents.

| i(in) | Rf | v(out), meas. | v(out), calc. | % Error |
|-------|-----|---------------|---------------|---------|
|       |     |               |               |         |
|       |     |               |               |         |
|       |     |               |               |         |
|       |     |               |               |         |
|       |     |               |               |         |

Question 13.  In which situation will the input offset voltage affect the
measurement of small currents more, when the current source is a
small voltage and small resistor or when it is a large voltage
and large resistor?

Locate a signal diode (1N914) in your loose parts supply and connect the
input circuit of figure 5-9c to measure the reverse bias current of the diode.
Select the value of Rf necessary to give a reasonable value of v(out).  Record
the current for reverse bias voltages of 1, 2 and 5 V.

| v(reverse) | v(out) | Rf | i(rev), calc. |
|------------|--------|-----|---------------|
| 1 V        |        |     |               |
| 2 V        |        |     |               |
| 5 V        |        |     |               |

## 5-5 INVERTING AND SUMMING AMPLIFIER

Objectives: To develop the current follower into the inverting and summing amplifiers and to introduce the operation of balancing the amplifier offset voltage by balancing an LF351 op amp and using it to test the responses of inverting and summing amplifier circuits.

As illustrated in section 5-4, the input offset voltage of op amps can contribute significantly to amplifier output errors. Many op amps such as the LF351 on the op amp job board have provision for adjusting the offset to zero. In preparation for subsequent experiments that require small offsets, the LF351 will now be balanced.

A simple procedure that works well for any op amp configuration is to connect the amplifier with a gain of approximately 10, connect the input to common, and adjust the balance potentiometer until the op amp output is nearly zero. By referring to figure 5-10 and the diagram of the op amp job board of figure 5-2, connect the LF351 as an inverting amplifier with Rf = 100 k$\Omega$ and R(in) = 10 k$\Omega$.

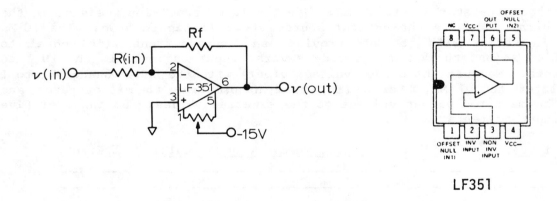

Fig. 5-10

Connect the free end of R(in) to common, and measure v(out) with the DMM on the most sensitive scale. Adjust the balance pot until v(out) is 1 mV or less (v(offset) $\leq$ 0.1 mV).

Now remove the common jumper to R(in) and use the VRS to provide the input voltages. Measure v(out) for five or more values of v(in) ranging from -0.7 to +0.7.

| v(in) | v(out), meas. | v(out), expected | % Error |
|-------|---------------|------------------|---------|
|       |               |                  |         |
|       |               |                  |         |
|       |               |                  |         |
|       |               |                  |         |
|       |               |                  |         |
|       |               |                  |         |

Based on the knowledge that the summing point of the op amp is at virtual common and based on previous experiments in this chapter, would you expect the inverting amplifier to load the VRS?  Why or why not?

Design an experiment to test your answer to the above question, execute the experiment, and discuss your results.

Change Rf to 10 MΩ and R(in) to 100 kΩ.  Apply five or more VRS voltages of both polarities and measure v(out).  Record the results below.

| v(in) | v(out), meas. | v(out), expected | % Error |
|-------|---------------|------------------|---------|
| _____ | _____ | _____ | _____ |
| _____ | _____ | _____ | _____ |
| _____ | _____ | _____ | _____ |
| _____ | _____ | _____ | _____ |
| _____ | _____ | _____ | _____ |
| _____ | _____ | _____ | _____ |

Once the LF351 has been carefully balanced, it should not require rebalancing within the length of a single laboratory period.  The balance should, however, be checked prior to each use, and adjusted if needed (if v(offset) > 0.5 mV).  A quick check can be made using the voltage follower configuration.

The inverting amplifier configuration is used to perform several mathematical operations. The summing amplifier provides an output related to the algebraic sum or two or more signals.

Connect the summing amplifier shown in figure 5-11.

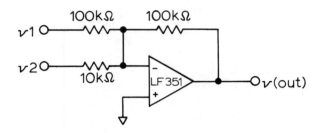

Fig. 5-11

Use the $\overset{+}{-}10$ V VRS output as voltage source v1 and the $\overset{+}{-}1$ V VRS output as source v2. Measure v(out) for six or more combinations of input voltage values. Keep one value constant for at least three values of the other and vice-versa.

| v1 | v2 | v(out) |
|----|----|--------|
| ____ | ____ | ____ |
| ____ | ____ | ____ |
| ____ | ____ | ____ |
| ____ | ____ | ____ |
| ____ | ____ | ____ |
| ____ | ____ | ____ |
| ____ | ____ | ____ |

Graph 1. Plot v(out) versus v2 for constant v1 and v(out) versus v1 for constant v2. Identify the slopes and intercepts of the plots and values of the circuit components.

Disconnect the summing network from the LF351 in preparation for the following experiment. Do not disconnect the balance pot or power connections to the LF351.

Question 14. Design and describe an inverting amplifier with a thermistor as one resistor such that the output voltage becomes more positive as the temperature increases. The thermistor resistance is 10.5 k$\Omega$ at 28$^{\circ}$C, and 9.5 k$\Omega$ at 23$^{\circ}$C. Choose the component values such that the op amp output changes 10 mV per $^{\circ}$C near room temperature. Also include an offset circuit so that the output voltage is 250 mV at 25$^{\circ}$C. Assume that the thermistor resistance is linearly related to the reciprocal of the absolute temperature under these conditions.

## 5-6  OP AMP INTEGRATOR

Objectives:  To observe the response and limitations of the analog integrator
by wiring a carefully balanced op amp and a good quality
capacitor into the integrating circuit and observing its response
to dc signals, step changes in signal level, and wave forms from
the FG.

When a capacitor is used as the feedback element in the inverting
configuation, the result is the op amp integrator such as the one illustrated
in figure 5-12.

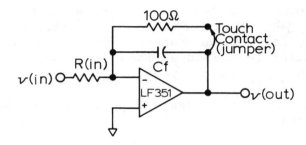

Fig. 5-12

Wire the LF351 op amp integrator as shown in figure 5-12.  Use a 1 µF low
leakage capacitor of 10% or better tolerance.  High quality capacitors are
necessary in this application since small deviations from ideality in the
capacitors may contribute substantial error.

With R(in) = 1 MΩ and v(in) = 100 mV, measure the times required for the
output to change by 1 volt, 3 volts, 5 volts, and 8 volts.  Begin the timing
when the touch contact is removed.  Repeat the  v(out) = 1 V measurement three
times and report the precision of the measurements.  The jumper can be used to
discharge the integrating capacitor through the 100 Ω resistor.

| v(out) | t, experiment | t, calc. | % error |
|--------|---------------|----------|---------|
|        |               |          |         |
|        |               |          |         |
|        |               |          |         |
|        |               |          |         |
|        |               |          |         |

The precision of single measurements is _____
(relative standard deviation).

Is the measurement   accurate   to   within   component   tolerances?
_____

Restart the integrator.  At v(out) ⁓ 1 V, rapidly remove the VRS signal
from the input, and observe v(out).  Does v(out) remain relatively constant?
_____.  If v(out) changes appreciably, rebalance the op amp.

76

Attach R(in) to common, reset the integrator, and observe v(out) on the most sensitive range of the DMM.

Observation:




Now the integrator will be used to measure the capacitance of an "unknown" capacitor. Make certain the op amp is balanced. Obtain an unknown capacitor, C(unk), and install the capacitor in the utility job board. Attach one end of the capacitor to common as shown in figure 5-13, and attach the other end to +5 V with a patch wire (position A).

Remove R(in) from the circuit of figure 5-12 to obtain the charge-to-voltage converter of figure 5-13.

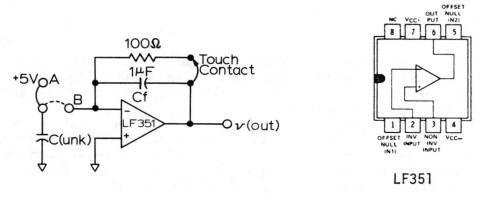

Fig. 5-13

Discharge Cf, carefully move the patch wire from position A to position B, and note the change in v(out). Repeat three to five times.

| Trial | v(out) |
|---|---|
| _____ | _____ |
| _____ | _____ |
| _____ | _____ |
| _____ | _____ |
| _____ | _____ |

v(out), mean = _____

Question 15. Calculate the unknown capacitance assuming Cf = 1.00 µF. Calculate the relative standard deviation of the measurement.

Question 16. Describe your observations and explain any deviations from ideal behavior.

Set up the circuit of figure 5-14.

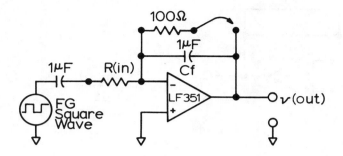

Fig. 5-14

The 1.0 µF capacitor between the FG and the integrator serves to remove the dc component of the square wave. If your FG has a dc offset control, adjust the offset to zero. Observe both the FG input and the integrator output on the scope.

Sketch the observed waveforms with appropriately labeled axes. Change R(in) to 10 kΩ and record any changes in the waveform.

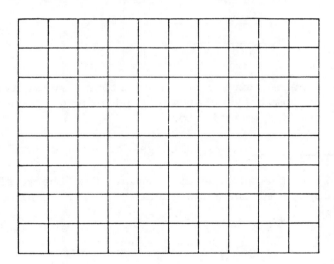

Question 17. Explain the shape of the observed waveforms.

## 5-7 OP AMP DIFFERENTIATOR

Objectives: To introduce the op amp differentiator by wiring the circuit and observing its response to various FG waveforms.

Reversing the positions of the resistor and capacitor of the op amp integrator produces the differentiator function. Connect the circuit of figure 5-15 using the component values shown and the TL082 op amp. Select a 1 kHz, 5 V p-p triangular wave as the signal source for the differentiator.

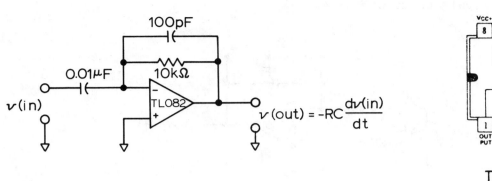 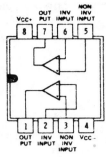

$$v(out) = -RC\frac{dv(in)}{dt}$$

TL082

Fig. 5-15

Sketch the input and output waveforms indicating the proper units for the ordinate and the abscissa.

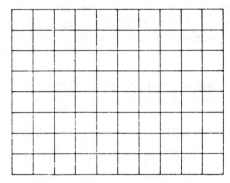

From the time base setting and the vertical sensitivity of the scope, calculate the slope of the triangular wave and record the height of the square wave resulting from the differentiation.

Slope = _____      Height of square wave = _____

Question 18. Calculate the theoretical value for the differentiator output and compare to the experimental value. Explain any differences.

Now change FG function to the square wave and sketch the resulting waveforms.

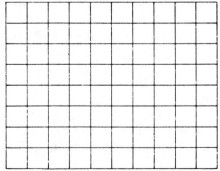

Question 19. Describe how a differentiator can be used to advantage in a particle counter (Coulter counter, Geiger counter, etc.).

## 5-8  OP AMP CHARACTERISTICS

Objectives:  To compare the limiting characteristics such as offset voltage
and input bias current among several types of commonly available
op amps by measuring these characteristics for each of three or
more types.

It is important to recognize the limitations of op amps so that
measurement errors may be avoided in instrumental applications.  We will
explore two important characteristics of several different common integrated
circuit op amps.

To measure the input offset voltage v(offset), connect the op amp as a
voltage follower and connect the non-inverting input to common.  The output
voltage equals v(offset).  Perform this measurement for the three different
types of op amps on the job board and include the data in the table.  Repeat
for at least two of the TL082 dual op amps.  Note that the pinout of the 741
op amp is identical to that of the LF351.

To measure the input bias current, i(bias), a 10 MΩ resistor should be
connected between the non-inverting input of the voltage follower and common.
The IR drop across the resistor results from the bias current.  The output
voltage v(out) is the sum of the offset voltage, v(offset), and the IR drop
across the resistor.  Thus i(bias) = [v(out) - v(offset)]/R.  Carry out this
determination for the op amps under investigation and enter the results in the
table.

| Op Amp | v(offset) | i(bias) |
|--------|-----------|---------|
| TL082  | _____ | _____ |
| LF351  | _____ | _____ |
| μA741  | _____ | _____ |

Question 20.  Op amp current-to-voltage converters are often used as readout
devices in photomultiplier circuits.  A particular application
requires the measurement of a current of $10^{-8}$ A.  Calculate the
measurement error due to input bias current for each op amp
studied above.

Question 21.  What effect does v(offset) have on the measurement described in
Question 20?

# UNIT 6. PROGRAMMABLE ANALOG SWITCHING

In this unit, several topics related to the switching of electronic signals are presented. The operating speed and contact bounce characteristics of an electromechanical relay are studied. Resistor-capacitor (RC) networks are examined in order to understand the influence of circuit time constants on the time response to pulsed signals. RC networks also form the basis of monostable and astable multivibrators. A TTL monostable circuit is examined and used to trigger a 555 integrated circuit timer. The two most common operational modes of the 555 timer are wired and tested. Both the AC and DC characteristics of an integrated circuit FET switch are determined. The switch is then used in a programmable gain amplifier, in an analog multiplexer, and in a digital-to-analog converter.

| Equipment | Parts |
|---|---|
| 1. Basic station | 1. 5% Resistors, 1 k$\Omega$(2), 10 k$\Omega$(3), |
| 2. CTFM | 100 k$\Omega$(2), 125 k$\Omega$, 250 k$\Omega$, 500 k$\Omega$, 1 M$\Omega$ |
| 3. Reference job board | 2. Capacitors, 220 pF(2), 0.01 $\mu$F(2) |
| 4. Utility job board | |
| 5. Op amp job board | |
| 6. BSI job board | |
| 7. Signal conditioning job board | |
| 8. Basic gates job board | |

## 6-1 RELAY SWITCHES

Objective: To investigate the switching speed and contact bounce characteristics of an electromechanical relay by observing the operate, transfer, bounce and break times on an oscilloscope.

Electromechanical relays are used in analog switching applications where very high off and very low on resistances are essential, or where high power requirements make solid state switches impractical. Disadvantages of electromechanical relays include relatively slow operation and contact bounce. These important characteristics are measured in this section.

Locate the relay and a 7406 hex TTL driver on the signal conditioning job board as shown in figure 6-1. The 7406 driver is capable of providing the current (25 mA) needed to actuate the relay coil. The output of the function generator, for example, would not be able to provide enough current for actuating the relay.

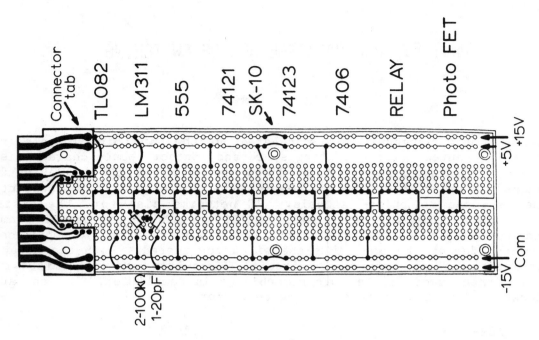

Fig. 6-1

Figure 6-2a shows a pictorial of the relay and the wiring diagram is shown in figure 6-2b. The pinout of the 7406 driver is shown in figure 6-2c.

The circuit of figure 6-2d is designed to produce 0 V on the moving contact when it is in the NC position, 5 V when it is in the NO position, and 2.5 V when in between. Wire the circuit in figure 6-2d. To verify that the circuit is working, touch the TTL input to common repeatedly, and listen carefully to hear the relay operate. Disconnect the input from common, and connect it instead to the function generator TTL output. Set the function generator frequency to about 100 Hz. Observe the function generator TTL output on scope Ch 1 and the voltage on the moving contact on scope Ch 2. The resulting waveform should be similar to figure 6-3.

Measure the operate, transfer, bounce, and break times on the scope. Perform the measurement at three different function generator frequencies in the 50 Hz to 150 Hz range. Also determine at which frequency the movable contact fails to make contact with the NO contact.

| FG freq. | Operate time | Transfer time (make) | Bounce time (make) | Break time | Transfer time (break) | Bounce time (break) |
|---|---|---|---|---|---|---|
| ____ | ____ | ____ | ____ | ____ | ____ | ____ |
| ____ | ____ | ____ | ____ | ____ | ____ | ____ |
| ____ | ____ | ____ | ____ | ____ | ____ | ____ |

Highest frequency for contact _____

Question 1. Which times in the above table, if any, change as a function of input frequency? Explain.

## Structure

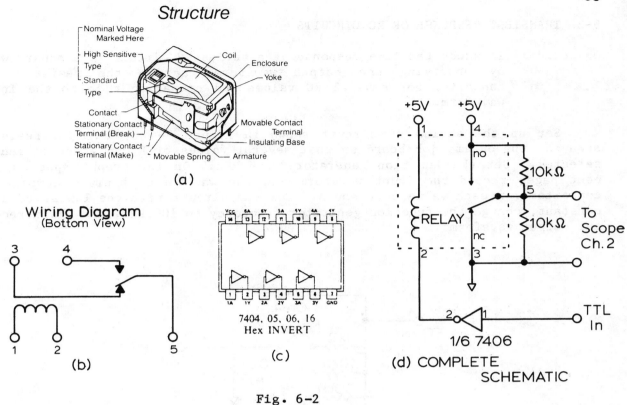

Fig. 6-2

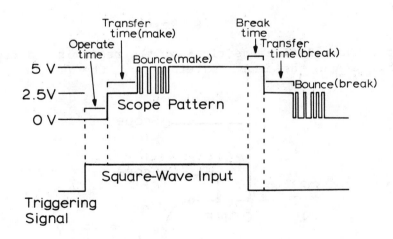

Fig. 6-3

## 6-2  TRANSIENT RESPONSE OF RC CIRCUITS

Objective:  To study the time response of a series RC circuit to a square wave
by observing the output waveforms across the resistor and
capacitor for several RC values and comparing them to the input
waveform.

Set up the series RC circuit shown in figure 6-4 by using a resistor
array on the op amp job board so that several values of R can be conveniently
selected.  Set the function generator to 1 kHz.  On the graph paper on the
next page, record the input waveform and the waveforms observed across the
capacitor for each value of R shown.  For the circuit with the lowest RC time
constant, change the function generator frequency to 100 kHz, and again record
the output waveforms.

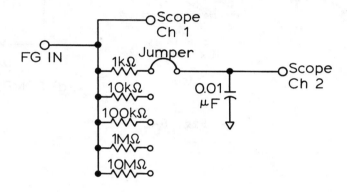

Fig. 6-4

Now reverse the relative positions of the resistor and capacitor as shown
in figure 6-5.  Graph the input waveform and the output taken across the
resistor.

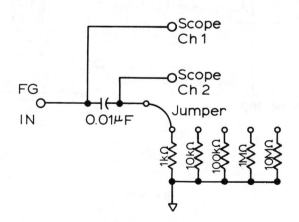

Fig. 6-5

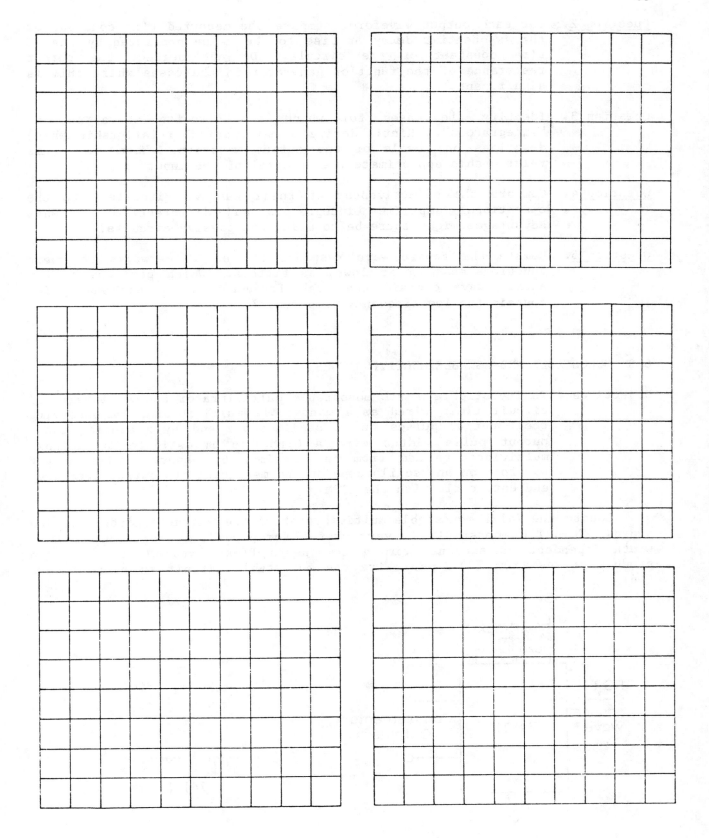

Question 2. For each output waveform, compare the measured time constant of the exponential decay or rise to the value predicted by the RC time constant of the circuit. Do not neglect the output resistance of the function generator in the cases where this is significant.

Question 3. Identify the cases for which the circuits above act as "integrators". Hint: derive a mathematical relationship which describes one cycle of the output waveform. When does this relationship approximate the integral of the input?

Question 4. Compare the effectiveness of these passive circuits with the operational amplifier integrators and differentiators. What advantages might there be to using the passive circuits?

Question 5. Relate the square wave response of the RC networks to their functions as high or low pass filters. Which portions of the square wave contain the high frequency components and which contain the low frequency components?

## 6-3 MONOSTABLE AND ASTABLE MULTIVIBRATORS

Objectives: To investigate a TTL monostable multivibrator and an integrated circuit timer wired as a monostable multivibrator by observing their output pulses on an oscilloscope and by measuring their output pulse widths with a CTFM; to investigate an astable multivibrator made from an IC timer by observing its output waveform on an oscilloscope and by measuring its output frequency and duty cycle with the CTFM.

   Locate the 74121 monostable multivibrator on the signal conditioning job board. This TTL monostable is very useful for providing short pulses of a width dependent on external timing components R(ext) and C(ext). It can produce pulses from ~ 0.2 μs. Wire the monostable circuit shown in figure 6-6.

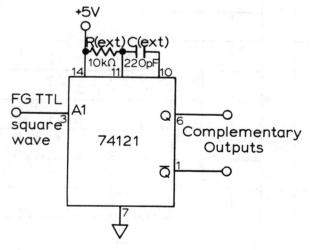
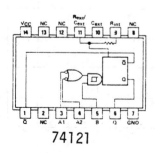

Fig. 6-6

Set the FG to produce an output frequency of ⁓ 1 kHz. Use the scope to observe the FG output (Ch 1) and the monostable output at Q and then Q̄. The output pulse should be 1-2 μs in duration and the scope intensity will have to be turned up to observe it. Record waveforms below and show the time relationship between the monostable trigger and the output pulse.

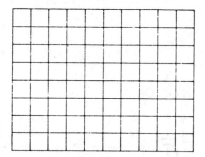

Is the pulse width dependent on the input frequency? _____

Now measure the output pulse width accurately by using the time A-B mode of the CTFM as shown in figure 6-7. The output at Q provides a falling edge to start the CTFM timer; the falling edge of the Q̄ output is used to stop the timer.

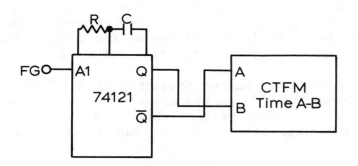

Fig. 6-7

Record the pulse width measured and the pulse width obtained with a second 220 pF capacitor in parallel with C(ext). Compare the two pulse widths with the theoretical values obtained from t(pw) = R(ext) C(ext) ln 2.

| C(ext) | R(ext) | Measured t(pw) | Theoretical t(pw) |
|--------|--------|----------------|-------------------|
| _____ | _____ | _____ | _____ |
| _____ | _____ | _____ | _____ |

Remove the second 220 pF capacitor so that C(ext) = 220 pF and R(ext) = 10 kΩ. The 74121 will now be used as a trigger input to the 555 integrated circuit timer. Locate the 555 on the signal conditioning job board. Wire the circuit shown in figure 6-8. The FG should still be connected to the 74121. Use the $\overline{Q}$ output of the 74121 as the trigger input to the 555.

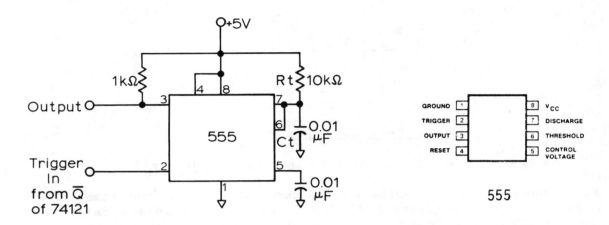

Fig. 6-8

The reason for using the 74121 to trigger the 555 is that the trigger operation on the 555 varies somewhat with manufacturer. Most 555's (from Signetics, Texas Instruments, Motorola, and others) change to the unstable state as soon as the trigger goes LO, but do not begin timing until the trigger goes back to HI. Others, especially those from Intersil, change state and begin timing on the HI to LO transition of the trigger input. The 74121 pulse generator assures proper triggering of the 555 regardless of manufacturer.

Set the FG to produce an output frequency of ~ 1 kHz. Observe the FG output and the 555 output (pin 3) on the scope. Sketch the waveforms observed showing the timing relationship between the FG output and the 555 output. Also sketch the relationship between the voltage on pins 6 and 7 and the 555 output.

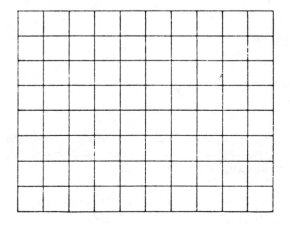

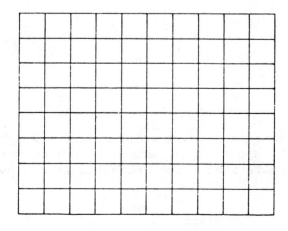

Question 6. In theory, the capacitor connected to pin 6 should charge to 2/3 of the 5 V supply. Determine from the sketch the time constant of the exponential rise observed at pin 6. Compare it to the value of Rt x Ct used.

Measure the pulse width of 555 timer using the time A-B mode of the CTFM as shown in figure 6-9.

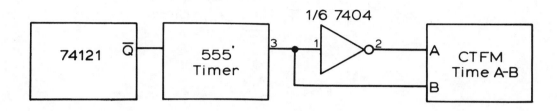

Fig. 6-9

The 7404 inverter provides a falling edge to start the CTFM timer and the 555 output falling edge stops the CTFM timer. Record the pulse width of the original circuit, and the pulse width obtained with a second 0.01 μF capacitor in parallel with the first. Make enough measurements to determine the precision of the pulse width.

| Ct | Rt | t(pw) |
|-----|-----|-----|
| ____ | ____ | ____ |
| ____ | ____ | ____ |
| ____ | ____ | ____ |
| ____ | ____ | ____ |
| ____ | ____ | ____ |
| ____ | ____ | ____ |

Question 7. Derive the theoretical below expression for the pulse width t(pw) = 1.1 Rt x Ct. Compare the measured and expected values. Account for any discrepancies that exist.

A 555 timer IC can be used as a very inexpensive oscillator. By connecting the TRIGGER input to the voltage across the timing capacitor, the discharging of the capacitor will start another charging cycle. Connect the circuit of figure 6-10. Be sure to remove the previous trigger circuit. The charging time constant is (Rt1 + Rt2) x Ct; the discharge time constant is Rt2 x Ct.

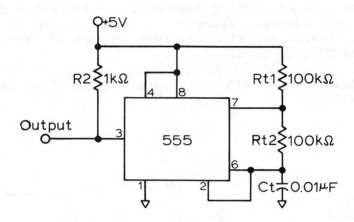

Fig. 6-10

Sketch the output waveform with sufficient detail to verify the charge and discharge time constants, and to determine the frequency of operation.

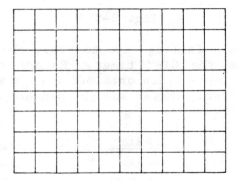

Use the time A-B mode of the CTFM to measure the duty cycle of the astable multivibrator accurately. Use the CTFM in the frequency mode to measure the frequency of the astable oscillator for the values of Ct and Rt shown.

| Ct | Rt1,Rt2 | Frequency |
|---|---|---|
| 0.01 µF | 100 kΩ | _____ |
| 0.1 µF | 100 kΩ | _____ |
| 1.0 µF | 100 kΩ | _____ |
| 0.01 µF | 10 kΩ | _____ |
| 0.1 µF | 10 kΩ | _____ |
| 1.0 µF | 10 kΩ | _____ |

Question 8. Compare the measured frequencies to those calculated from the following equation:

$$f = \frac{1.44}{(Rt1 + 2Rt2)Ct}$$

Disconnect this circuit.

## 6-4 CHARACTERISTICS OF SOLID STATE SWITCHES

Objective: To study the switching characteristics of a solid state analog switch by measuring the switch ON and OFF resistances, the leakage current and the turn on and turn off times.

Locate the LF13202 quad single pole-single throw (SPST) FET analog switch on the operational amplifier job board. The pinout is given in figure 6-11. When using the LF13202, be certain to turn on the power to the job board before applying an external signal to the IC.

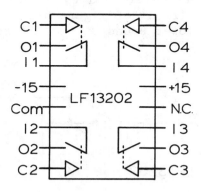

Fig. 6-11

This switch can transmit current or voltage signals from the input (I) to the output (O). The condition of the switch (open or closed) depends upon the logic level (LO or HI) applied to the control input (C). When the control input is HI, the switch is closed, and when the control input is LO or unconnected, the switch is open.

The ON resistance of the analog switch can be measured by applying a constant current through a closed switch and measuring the voltage drop developed across the switch. A voltage follower is used to isolate the switch from the DMM. Note that the voltage follower is a _balanced_ LF351.

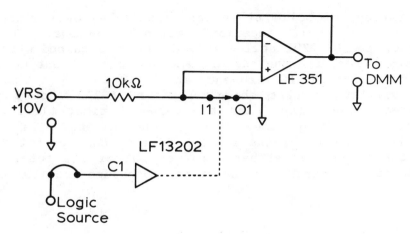

Fig. 6-12

Wire the circuit in Figure 6-12. Use analog switch 1. MAKE ABSOLUTELY CERTAIN THE POWER AND COMMON CONNNECTORS ARE CORRECTLY WIRED!! Use a logic level source to turn the switch on and measure the IR drop across the switch, V(sw).

V(sw) _____

<u>Question 9</u>. Calculate the switch ON resistance. Suggest a measurement situation in which an ON resistance of this magnitude could cause an error.

Now the OFF resistance is determined by measuring the leakage current in the OFF state. Connect the circuit of figure 6-13. Record vo below. You may be able only to establish a lower bound on the OFF resistance.

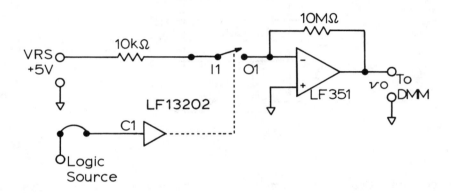

Fig. 6-13

vo = _____

<u>Question 10</u>. Calculate the leakage current and the OFF resistance.

<u>Question 11</u>. Why was the LF351 used in this experiment?

A major advantage of solid state analog switches over electromechanical and reed relays is their superior switching speeds. The switching characteristics of the LF13202 analog switch can be obtained with the circuit shown in figure 6-14a. Wire the circuit and use the TTL output of the FG as the control signal. Set the FG frequency at 10 kHz.

A display similar to that shown in figure 6-14b should be obtained. Record the turn-on time, t(on), and the turn-off time, t(off). Note that t(on) is composed of a delay time td and a rise time tr. The delay time is the time from the occurence of the leading edge of the control signal to the time at which the output signal has achieved 10% of its total change. The rise time is the time required for the signal to change from 10% to 90% of its total change.

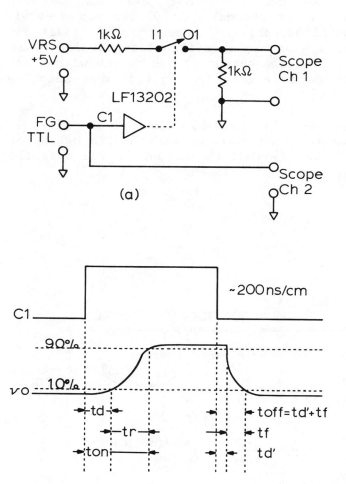

Fig. 6-14

Upon turn-off, the delay time is the time from occurence of the falling edge of the control signal to the time at which the signal has dropped to 90% of its full on value. The fall time is the time required for the signal to drop from 90% to 10% of its full on value. It is not unusual for the delay time or turn-off to be immeasurably small. Record the important times below.

td, on  = _____     td, off = _____
tr      = _____     tf      = _____
t(on)   = _____     t(off)  = _____

<u>Question 12</u>.  Compare relays and analog switches.  Cite situations for which each is uniquely suited.

## 6-5  ANALOG MULTIPLEXER

Objectives: To study the use of analog switches in multiplexing by constructing a two-channel multiplexer and observing its output on the oscilloscope; to measure the isolation between two switches on a quad analog switch integrated circuit by measuring the ac signal at the output of one switch with a dc signal at its input while an ac signal is applied to a second switch on the same IC.

Wire the circuit shown in figure 6-15.  Connect a 5 V p-p 1 kHz sine wave from the function generator to one switch, and the TTL output of the function generator to the other.  Verify that the output of the operational amplifier depends upon which analog switch is closed.

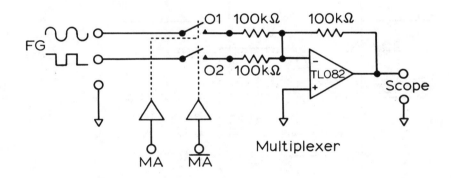

Fig. 6-15

Ideally, only the signal connected to the closed switch should be present at the output of the operational amplifier.  In fact, a certain amount of cross-talk will occur.  This can be measured by connecting a constant voltage to one switch input, and the 1 kHz, 5 V p-p ac sine wave to the other.  With the constant voltage selected as the input to the multiplexer, measure the amplitude of the 1 kHz ac component of the output.

Amplitude, p-p = _____

Question 13.  Express the amount of cross-talk as a ratio in dB.  Give some situations in which cross-talk might cause a problem.

6-6  PROGRAMMABLE GAIN AMPLIFIER AND DIGITAL-TO-ANALOG CONVERTER

Objectives:  To investigate a programmable gain amplifier constructed from an operational amplifier and analog switches; to select the divider network for a follower-with-gain by measuring the output voltage that results when different switches are closed; to study a binary digital-to-analog converter by obtaining the output voltage for several digital input numbers.

Wire the programmable gain amplifier circuit of figure 6-16. Use a balanced LF351 operational amplifier and three of the LF13202 analog switches. Connect the control inputs of the switches to three logic sources. To avoid strange gain combinations, only one switch should be closed at a time. Verify the operation of the circuit by making the measurements called for in the table below. Use the VRS - 1 V output for v(in).

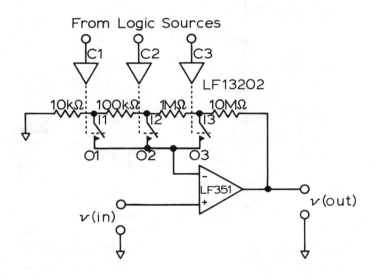

Fig. 6-16

| v(in) | switch closed | v(out) | Gain |
|-------|---------------|--------|------|
| 10.0 mV | 1 | _____ | _____ |
| 100.0 mV | 1 | _____ | _____ |
| 10.0 mV | 2 | _____ | _____ |
| 100.0 mV | 2 | _____ | _____ |
| 10.0 mV | 3 | _____ | _____ |

Question 14.  Compare the calculated gain and the measured gain for each switch setting. Account for any discrepancies.

Question 15.  What effect, if any, does the switch ON resistance have?

The digital-to-analog converter of figure 6-17 is a type of programmable gain amplifier in which the analog input voltage is "amplified" by an amount determined by which of the analog switches are closed. The current added to the summing point of the operational amplifier by the closing of a switch is proportional to the digit value of that switch in the number system under which the analog-to-digital converter operates. Wire the circuit of figure 6-17. Obtain the -10 V from the voltage reference source. The weighting of the resistors used in this circuit is binary. In the table below, record the output voltage for each possible binary input (0000 to 1111).

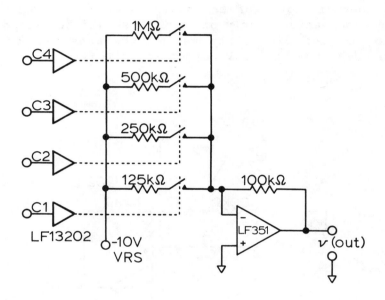

Fig. 6-17

| Input number | Measured v(out) |
|---|---|
| _____ | _____ |
| _____ | _____ |
| _____ | _____ |
| _____ | _____ |
| _____ | _____ |
| _____ | _____ |
| _____ | _____ |
| _____ | _____ |
| _____ | _____ |
| _____ | _____ |
| _____ | _____ |
| _____ | _____ |
| _____ | _____ |
| _____ | _____ |
| _____ | _____ |
| _____ | _____ |

Question 16. Compare the measured v(out) values to the expected values. Account for any errors.

   Now connect a 1 V p-p 1 kHz sine wave from the function generator in place of the voltage reference source. Record the p-p output amplitude for each number input.

| Number input | p-p amplitude |
| --- | --- |
| _____ | _____ |
| _____ | _____ |
| _____ | _____ |
| _____ | _____ |
| _____ | _____ |
| _____ | _____ |
| _____ | _____ |
| _____ | _____ |
| _____ | _____ |
| _____ | _____ |
| _____ | _____ |
| _____ | _____ |
| _____ | _____ |
| _____ | _____ |
| _____ | _____ |
| _____ | _____ |

Question 17.   Why is this circuit often called a multiplying digital-to-analog converter?  Suggest an application for a multiplying DAC.

# UNIT 7. SOLID STATE SWITCHES AND AMPLIFIERS

In this unit the characteristic curves of a bipolar junction transistor (BJT), a field effect transistor (FET), a silicon controlled rectifier (SCR) and a triac are obtained and used to determine important device characteristics. The switching speed of a BJT circuit illustrates the high frequency limitations of these devices. A basic transistor amplifier is wired and its gain and phase shift are measured as a function of frequency. Several other amplifier circuits are studied including an emitter-follower, a difference amplifier and an operational amplifier.

<u>Equipment</u>

1. Basic station
2. Characteristic curve tracer with instruction manual
3. Utility job board
4. Op amp job board
5. Power supply job board
6. Signal conditioning job board

<u>Parts</u>

1. 2N2222 Transistor
2. 2N5459 FET
3. 2N3958 matched JFET pair
4. SCR (5-10A, 100-200V)
5. Triac (5-10A, 100-200V)
6. 5% resistors: 7.5 Ω, 56 Ω, 100 Ω, 680 Ω, 470 Ω, 1 kΩ, 4.7 kΩ, 10 kΩ, 22 kΩ, 47 kΩ, 100 kΩ (5), 220 kΩ
7. Capacitors: 47 pF, 68 pF, 0.1 μF (2), 1 μF
8. Lamp, CM53
9. Phototransistor (FPT-100, Radio Shack)
10. Potentiometer, 100 kΩ
11. Optically-isolated, zero crossing switch, Electrol SA-8

## 7-1 BIPOLAR JUNCTION TRANSISTOR CHARACTERISTICS

Objectives: To obtain the characteristic curves of a BJT, to measure the dc and ac current gains ($\beta$'s), and to observe the saturation characteristics $V(CE)$,sat and $R(ON)$.

A commercial curve tracer is necessary for this experiment. Consult the instruction manual for the details of its operation.

Obtain a 2N2222 NPN small signal transistor and insert it into the socket provided on your curve tracer. Be sure to observe the proper location of the emitter, base, and collector as shown in the pin diagram of figure 7-1.

Fig. 7-1

Adjust the curve tracer and/or oscilloscope controls to correspond to the display conditions listed below:

| | |
|---|---|
| Horizontal Display, V(CE) | 10 V full scale |
| Curve tracer Sweep Range | 0-10 V |
| Vertical Display, IC | 5-20 mA full scale |
| Step Size, IB | 0.001-0.01 mA/step |
| Limiting Resistance | 500 Ω - 5 kΩ |

Vary the limiting resistor over the range shown, and note the effect on the slope of a line connecting the ends of the curves. This line is parallel to the load line for the same resistance value as shown in figure 7-2. To obtain a load line for a particular circuit, construct a line connecting points IC = V(CE)/R and V(CC) = V(CE).

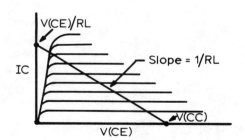

Fig. 7-2

Sketch or photograph a family of collector characteristic curves, carefully label the plots and draw a load line for V(CE) = 5V and RL = 1 kΩ.

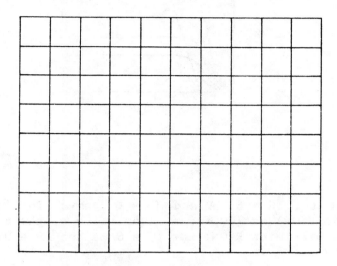

The dc current gain, β(dc) will now be measured. Decrease the size of the limiting resistor to ˜ 100 Ω, expand the vertical display to correspond to ˜ 50 mA full scale, and change the step size to ˜ 0.05 mA/step. Obtain IC at V(CC) = 5V and IB = ˜ 0.1 mA. Calculate β(dc). Repeat for other values of V(CE) in the range of 3V to 7V and other values of IB in the range of 0.05 mA – 0.2 mA. Record the results in the table below.

| V(CC) | IB | IC | β(dc) |
|-------|------|------|-------|
| 5V | 0.1 mA | _____ | _____ |
| _____ | _____ | _____ | _____ |
| _____ | _____ | _____ | _____ |
| _____ | _____ | _____ | _____ |

Question 1. Does β depend on the values chosen for V(CC) and IB? Explain.

The ac current gain, β(ac) = ΔIC/ΔIB, can be obtained from the family of curves just observed. Measure the difference in collector current, ΔIC, between two base current steps at V(CE) = 5V and record the results below.

| ΔIC | ΔIB | β(ac) |
|------|------|-------|
| _____ | _____ | _____ |

Question 2. Is β(dc) significantly different from β(ac)? Explain.

Next V(CE),sat will be measured. To display the saturation region, set the parameters to 1 V full scale horizontal sensitivity and 10 mA full scale vertically. The saturation region is the area at low V(CE) to the left of the linear region as shown by the dashed line of figure 7-3.

102

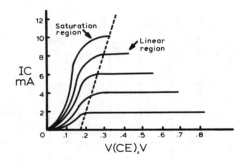

Fig. 7-3

Determine V(CE),sat for IC = 6 mA and IB = 0.1 mA. The "on" or saturation resistance, R(ON) is V(CE),sat/IC for a given value of the base current in the saturation region. Determine R(ON) for IC = 6 mA and IB = 0.1 mA.

V(CE),sat _____
R(ON) _____

7-2  FIELD-EFFECT TRANSISTOR CHARACTERISTICS

Objectives:  To obtain the characteristic curve of a junction field-effect transistor (JFET), and to measure the pinchoff voltage and transconductance of the FET.

Consult the curve tracer instruction manual for procedures to display FET characteristic curves. Insert a 2N5459 (or similar) FET into the socket on your curve tracer with the proper polarity according to the pin diagram of figure 7-4.

1 2 3

(Bottom View)

1 = drain
2 = source
3 = gate

Fig. 7-4

Arrange the curve tracer settings to obtain a family of curves of drain current, ID versus drain-to-source voltage, V(DS) for various values of gate-to-source voltage, V(GS). Adjust the sweep range so that it covers 0 to 20 V. Vary the limiting resistor and note the effect on the curves. Obtain the pinchoff voltage, VP, by noting the value of V(GS) which returns ID to approximately zero.

Sketch or photograph the display, carefully label the plot, and indicate VP.

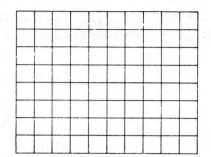

VP _____

Next the transconductance gm = $\Delta$ID/$\Delta$V(GS) will be measured.  Calculate gm values for several $\Delta$V(GS) values in the region of V(DS) = 5-10 V.

| V(DS) | $\Delta$ID | $\Delta$V(GS) | gm |
|-------|------|--------|----|
| ___ | ___ | ___ | ___ |
| ___ | ___ | ___ | ___ |
| ___ | ___ | ___ | ___ |
| ___ | ___ | ___ | ___ |
| ___ | ___ | ___ | ___ |
| ___ | ___ | ___ | ___ |

If specifications are available for your FET, determine whether the measured gm values are within the acceptable range given by the manufacturer.

Manufacturer's Specifications

| gm (min) | gm (typical) | gm (max) | Measured range of gm (values) |
|----------|--------------|----------|-------------------------------|
| ___ | ___ | ___ | ___ |

Question 3.  Describe why pinchoff occurs in FET devices.

Estimate the drain-source dynamic resistance, r(ds), of your JFET from the reciprocal of the slope of the ID vs. V(DS) curve at an intermediate value of V(GS).

r(ds)$\cong$ _____

Mark or otherwise identify the FET you have just characterized.  The values measured will be used in a later experiment.

7-3  THE SCR AND TRIAC

Objectives:  To obtain current-voltage curves for a SCR and a triac, to investigate the forward and reverse breakdown regions and to obtain the turn-on characteristics for triggering thyristors into conduction.

Consult the instruction manual for your specific curve tracer for details of the procedure for displaying characteristics of SCR's and triacs.

A.  SCR Characteristics

Obtain an SCR with the following characteristics (ON current ≅ 5-10 A, peak OFF voltage ≅ 100-200 V).  Attach the SCR to the curve tracer as shown in figure 7-5.

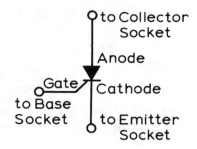

Fig. 7-5

Set up the curve tracer to measure the forward blocking voltage (breakover voltage) and holding current as follows:

> Horizontal sensitivity   ≅   200 V full scale
> Vertical sensitivity   ≅   30-50 mA full scale
> Curve tracer sweep range   ≅   maximum
> Gate current step size   =   OFF
> Sweep polarity   =   NPN
> Limiting resistance   ≅   10 kΩ

The resulting display should be similar to that shown in figure 7-6.  If no forward current is observed, your SCR has a breakover voltage that exceeds the sweep range of the curve tracer.  From the current-voltage curve measure the breakover voltage V(BO) and the holding current IH.

> V(BO)  =  _____
>
> IH     =  _____

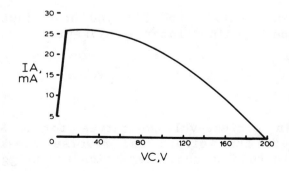

Fig. 7-6

<u>Question 4</u>.  What is the significance of the holding current?  Why is it important to know the breakover voltage?

Measure the reverse breakdown voltage V(RB) and reverse leakage current IR by reversing the curve tracer sweep polarity (change to PNP).  Again if no conduction is observed your curve tracer sweep range is less than the breakdown voltage.

$$V(RB) = \underline{\hspace{5cm}}$$

$$IR = \underline{\hspace{5cm}}$$

The SCR can be triggered into conduction (in the forward direction) at voltages below the forward breakover voltage by applying a small trigger current to the gate.  Set the curve tracer back to NPN polarity.  Consult the instruction manual for how to apply a single gate current step to the SCR (Set the base current step generator to apply 1 step).  Adjust the step size for the minimum gate current.  Set the other curve tracer parameters as below:

Horizontal sensitivity $\cong$ 50 V full scale
Vertical sensitivity $\cong$ 1 A full scale
Sweep polarity = NPN
Sweep range $\cong$ 0-50 V
No. of steps = 1
Gate current step size = minimum
Limiting resistor $\cong$ 50$\Omega$

Now increase the gate current step size until the SCR just goes into conduction, and the display is similar to that shown in figure 7-7.

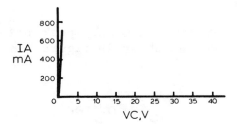

Fig. 7-7

The current required to trigger the SCR IT can be estimated from the gate current step size.

$$IT = \underline{\hspace{5cm}}$$

Sketch or photograph the current-voltage curve and carefully label the axes. Indicate on your graph the approximate minimum trigger current.

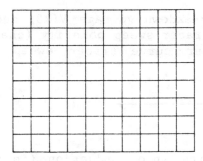

Now expand the scale of your display so as to observe the forward conduction in detail. Note that when the SCR conducts, there is a forward voltage drop across the device. Determine the forward voltage drop Vf at several anode-cathode currents IA in the range of 0-500 mA. Report your results below.

| IA | Vf |
|---|---|
| ——— | ——— |
| ——— | ——— |
| ——— | ——— |
| ——— | ——— |
| ——— | ——— |

Question 5.  How does the forward conduction characteristic of the SCR compare to that of a forward-biased diode?

Remove the SCR from the curve tracer.

B.  Triac Characteristics

Obtain a triac (ON current ≅ 5-10 A, peak off voltage 100-200 V) and connect it to the curve tracer as shown in figure 7-8.

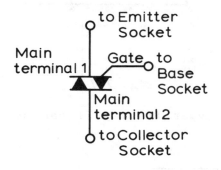

Fig. 7-8

The triac can be tested similarly to the SCR except that forward conduction can be tested in both directions. The triac can be triggered into conduction by either a positive or negative gate current.

Obtain the breakover characteristics first for quadrant I (Voltage at MT2 positive with respect to MT1 ) and then for quadrant III (V(MT2) negative). To reverse polarities merely change the curve tracer from NPN (quadrant I) to PNP (quadrant III).

$$V(BO),I \quad = \underline{\hspace{4cm}}$$

$$V(BO),III = \underline{\hspace{4cm}}$$

$$IH,I \quad\quad = \underline{\hspace{4cm}}$$

$$IH,III \quad\; = \underline{\hspace{4cm}}$$

Next obtain the minimum gate current to trigger the triac in both directions. If available on your curve tracer, use both + and – gate currents in both quadrants.

$$I(+),I \quad = \underline{\hspace{4cm}}$$

$$I(+),III = \underline{\hspace{4cm}}$$

$$I(-),I \quad = \underline{\hspace{4cm}}$$

$$I(-),III = \underline{\hspace{4cm}}$$

Investigate in detail the shape of the forward conduction characteristic in quadrants I and III.

Sketch or photograph the expanded i-V curve for the triac in both quadrants. Combine the curves/or both quadrants on the same graph.

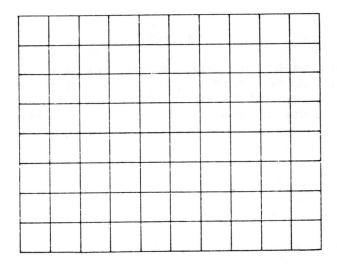

## 7-4  SCR AND TRIAC APPLICATIONS

Objectives:  To investigate further the useful characteristics of power
control devices by constructing and observing the behavior of an
SCR light activated circuit, a tungsten lamp ac controller, and
an optically isolated zero-crossing switch.

### A.  SCR Application

The circuit of figure 7-9 can be used to turn off the power to a load
(the lamp in this case) when the light intensity reaches a given level.  The
op amp integrates the photocurrent resulting from the light falling on the
phototransistor.  The sensitivity of the circuit is controlled by the value of
R, which discharges the feedback capacitor at a controlled rate.  Connect the
circuit as shown.  Begin with R ≅ 10 kΩ, and adjust the value up or down
depending upon ambient light levels in your laboratory.

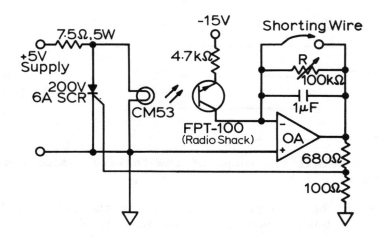

Fig. 7-9

Start the experiment with the power supply off and the light bulb well
separated from the phototransistor.  Apply power and gradually move the bulb
toward the sensor.  During initial operation it may be helpful to observe the
op amp output on the scope or DMM.  Describe your observations below.

Observations:

If your circuit worked properly, the SCR fired, and in order to be "rearmed", the power must be shut off for a few seconds. Rearm the circuit and gradually adjust R until the trigger distance is about 10 cm. Now measure the gate voltage (output of the voltage divider) at the trigger point for several trials, and record the results below. Is the trigger reproducible?

_____

<u>Divider voltage</u>

_____

_____

_____

_____

_____

<u>Question 6</u>. What is the function of the voltage divider at the op amp output?

B. Triac Application

The simple circuit of figure 7-10 can be used to control ac power to a tungsten filament lamp. Wire the circuit using the 14.1 V transformer used in Unit 3 as the power source. (Recall that the power supply job must be positioned in the rightmost frame slot to access the transformer secondary).

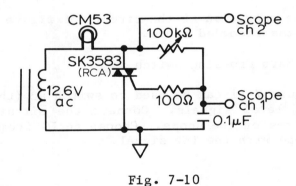

Fig. 7-10

Gradually adjust the potentiometer and note the effect on the lamp.

Observation:




Connect both channels of the scope to the circuit as shown in the figure.

110

Sketch pairs of waveforms for several different light intensity settings.

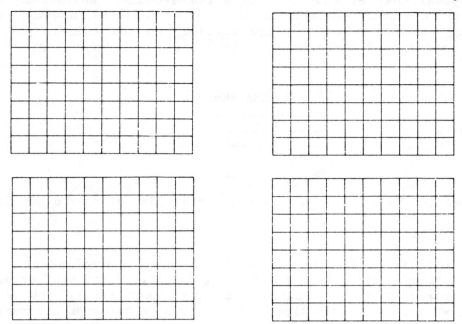

Question 7. Explain the operation of the circuit of figure 7-10 by discussing the waveforms recorded above.

C. Optically Isolated Zero Crossing Switch

The circuit of figure 7-11 can be used to switch ac (the FG in this case) signals with TTL logic level signals. Connect the circuit and display the input and output waveforms on the scope. Choose an FG frequency of about 500 Hz, and trigger the scope with the TTL signal.

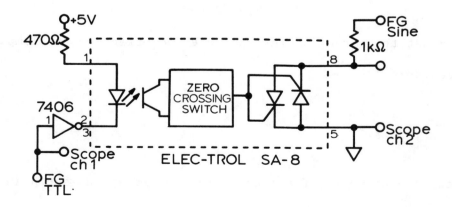

Fig. 7-11

Sketch pairs of waveforms for four or five different settings of FG offset.

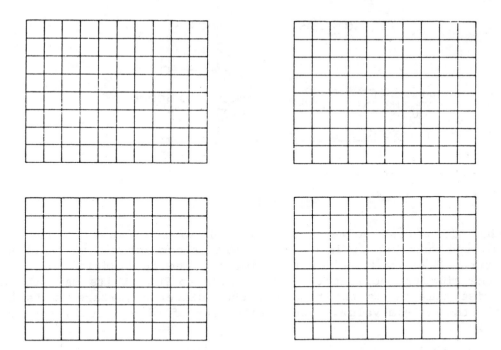

Question 8.    How does the IC zero crossing switch work?  What are the advantages of this type of switching?  Disadvantages?

## 7-5   TRANSISTOR SWITCHING SPEEDS

Objectives:  To determine turn-ON and turn-OFF times for a BJT switching circuit with and without saturation; to determine the effect of a "speed-up" capacitor in the base circuit.

Wire the transistor switching circuit of figure 7-12 leaving out the capacitor shown in dotted lines.  Set the FG frequency to $\cong$ 100 kHz and adjust the amplitude of the square wave to a voltage just below the onset of saturation.  (Saturation is readily observed because the BJT output voltage no longer increases with increasing input voltage).  Adjust the FG frequency, if necessary, so that the BJT output voltage completely reaches its minimum value before the next cycle of the input.

112

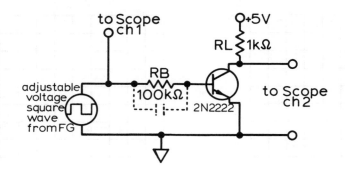

Fig. 7-12

Note that the BJT output voltage is inverted so that when ON the output voltage is LO, and when OFF it is HI, etc. With a non-saturating input signal measure the turn-ON delay time td; the time between application of the base voltage and the time the collector voltage falls to 90% of its maximum value. Measure the rise time tr as the time for the collector voltage to fall from 90% to 10% of its maximum value. The combination of td and tr is the turn-ON time t(on).

td    = _____

tr    = _____

t(on) = _____

Repeat for the delay time on turn-OFF, called the storage delay time ts and the fall time tf. If may be difficult to observe the storage delay time.

ts    = _____

tf    = _____

t(off) = _____

Sketch or photograph the waveforms observed.

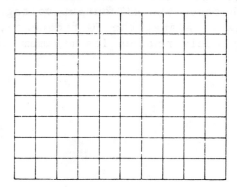
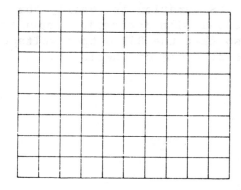

Question 9.  What causes the turn—ON delay time?  The storage delay time?
            Explain why the turn—ON and turn—OFF times differ.

     Now adjust the FG square wave amplitude so that the transistor is heavily
saturated (do not exceed 10 V p-p).  Observe the effect of increasing
saturation on the turn—ON and turn—OFF times.  Record your results.

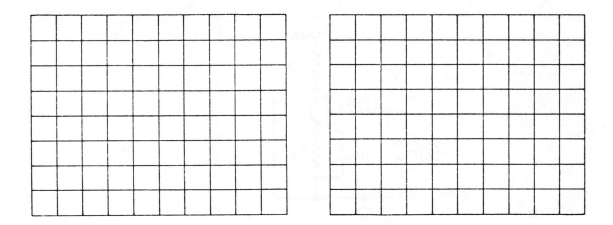

Question 10.  Describe why saturation changes the switching times.  Which time
             (ON or OFF) is most affected?  Why?

     Now the effect of a small "speed-up" capacitor in the base circuit will
be noted.  Place a capacitor (47 to 68 pF) in parallel with base resistor RB.
Arrange a touch contact so that its influence can be readily observed.
Measure the turn—ON and turn—OFF times with the "speed-up" capacitor for
saturating and non-saturating input voltages.  Compare the values obtained to
those measured without the capacitor.

|  | No Saturation | | Saturation | |
|---|---|---|---|---|
|  | with C | without C | with C | without C |
| td | _____ | _____ | _____ | _____ |
| tr | _____ | _____ | _____ | _____ |
| t(on) | _____ | _____ | _____ | _____ |
| ts | _____ | _____ | _____ | _____ |
| tf | _____ | _____ | _____ | _____ |
| t(off) | _____ | _____ | _____ | _____ |

Question 11.  Describe the effect of the "speed-up" capacitor and explain why
             it improves switching times.

114

## 7-6   BASIC TRANSISTOR AMPLIFIERS

Objectives:   To design a JFET amplifier with preselected gain, to develop a
self-biasing circuit and to measure its gain and phase shift as a
function of frequency.

The same FET that was characterized in experiment 7-2 will be used in
this experiment.  Refer to the characteristic curves obtained for the FET.  It
is desired to obtain a voltage gain Av ≅ 10.  The FET will be powered from the
+15 V supply as shown in figure 7-13.

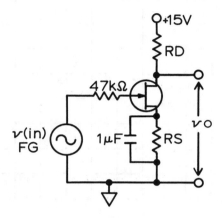

Fig. 7-13

From the measured transconductance gm for your FET choose an appropriate drain
resistor for a gain of ~ 10.

chosen RD = _____

From the characteristic curves choose an appropriate quiescent point to
put the FET in the middle of its linear range. Report the chosen quiescent
values for ID, V(GS) and V(DS).

ID    = _____
V(GS) = _____
V(DS) = _____

Resketch the characteristic curves below for your JFET amplifier with the
chosen load line and the Q point identified clearly.

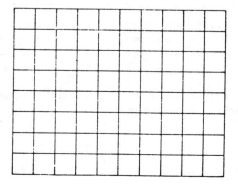

Determine the value of RS that will self-bias the JFET to produce the appropriate V(GS).

RS = _____

Before wiring the JFET amplifier circuit of figure 7-13, adjust the FG output so that it has no dc offset (if applicable). Select the square wave output, and a frequency of ~ 10kHz. Adjust the p-p output voltage to ~ 0.5 V and turn the FG off. Wire the JFET amplifier of figure 7-13 with your chosen component values. Include the 47 kΩ input resistor and the 1 µF bypass capacitor. Use the scope to observe the input waveform and output waveform. Calculate the gain. Be certain the JFET is not saturated. If the gain is greatly different from that expected (> 50%) you may have to adjust RD and the Q point.

Gain, Av = _____

Next the gain and phase shift will be obtained as a function of input frequency by means of Lissajous figures. Set the scope to obtain an X-Y display of vo vs v(in). Adjust the scope sensitivity to obtain a line of 45° angle for a phase shift of 180° (you may adjust the generator frequency until the phase shift is appropriate). Now vary the generator frequency in the range of 50 Hz – 1 MHz, and accurately measure the gain and phase shift for the frequencies in each decade listed below. (See page 44 of the text for obtaining phase shifts from Lissajous figures).

| f(in) | Av | φ |
|-------|------|------|
| 100 Hz | _____ | _____ |
| 200 Hz | _____ | _____ |
| 500 Hz | _____ | _____ |
| 1 kHz | _____ | _____ |
| 20 kHz | _____ | _____ |
| 5 kHz | _____ | _____ |
| 10 kHz | _____ | _____ |
| 20 kHz | _____ | _____ |
| 50 kHz | _____ | _____ |
| 100 kHz | _____ | _____ |
| 200 kHz | _____ | _____ |
| 500 kHz | _____ | _____ |
| 1 MHz | _____ | _____ |

Note: Some of the measured phase shift can come from the slope at these frequencies.

Graph 1. Plot the gain vs. frequency and the phase shift vs. frequency on log-log scales. Determine from your plot the upper and lower 3-dB points of the JFET amplifier.

Question 12. Compare the midfrequency gain with that calculated from the simple circuit model. Why might it be in error? What causes the decrease in gain at high and low frequencies? How could the frequency response of your JFET amplifier be improved?

Disconnect this circuit, but keep the FET available for use in the next experiment.

7-7  JFET SOURCE FOLLOWER

Objectives:  To wire a JFET source follower, to obtain its gain and phase
              shift, and calculate its input and output resistance.

        Wire the source follower of figure 7-14.  Use the FG sine wave output
(1-5 V pp) as the input signal.  Use the Lissajous figure technique as before
to observe the gain and phase shift characteristics.  Report the frequency at
which the high frequency gain deviates by 3-dB from the midfrequency gain.

        upper 3-dB point _____

Try to note any deviations at low frequency.

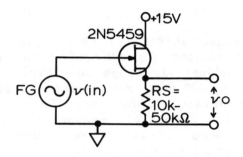

Fig. 7-14

        Use the values of gm and r(ds) obtained from the characteristic curves
and the value of Rs used to calculate the expected voltage gain Av.  Compare
to the observed gain.

|     | Calculated | Observed |
|-----|------------|----------|
| Av  | _____ | _____ |

Question 13.  Calculate from measured parameters (gm, r(ds), etc.), the input
              and output resistance of the source follower.  Compare the
              values obtained with Ro and R(in) for the common source
              amplifier of the previous experiment.  Show your calculations.


7-8  JFET DIFFERENCE AMPLIFIER

Objectives:  To wire a difference amplifier from matched JFETs, to obtain its
             dc and ac difference and common mode gains and its dc and ac
             common mode rejection ratios.

        Wire the circuit of figure 7-15, replacing the 0.1 μF capacitors shown in
dotted lines with direct connections.  Connect both inputs to common and
measure the differential output voltage (v(02)-v(01)).

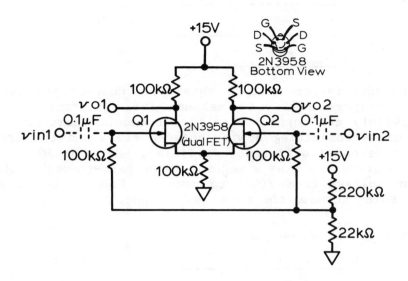

Fig. 7-15

Be sure the DMM is not attached to the frame system common when making this differential measurement. Any voltage measured is an offset voltage, V(os).

$$V(os) = \underline{\hspace{4cm}}$$

Now measure the dc difference signal gain by connecting v(in1) to the VRS and v(in2) to +5 V. Adjust the VRS voltage until the differential output voltage is in the range of +2 to +5 V. Measure the differential output voltage and the separate input voltages v(in1) and v(in2). The dc difference gain Ad is

$$Ad = \frac{v(02) - v(01)}{v(in1) - v(in2)}$$

vo = (v(02)-v(01)) = \underline{\hspace{4cm}}

v(in1)　　　　　 = \underline{\hspace{4cm}}

v(in2)　　　　　 = \underline{\hspace{4cm}}

Ad　　　　　　　 = \underline{\hspace{4cm}}

Now connect both inputs to +5 V and measure the common mode output voltage. Subtract the offset voltage. Calculate the dc common mode gain A(cm) = vo/v(cm) and the dc CMRR.

vo                = _____

V(os)             = _____

Corrected vo      = _____

A(cm)             = _____

CMRR              = _____

Now the ac characteristics will be obtained. Connect the 0.1 μF input capacitors shown in figure 7-15. To measure the difference signal gain, connect the FG to v(in1) and connect v(in2) to common. Short out the 100 kΩ drain resistor of Q2 by connecting a wire between +15 V and the drain of Q2. This produces a single-ended difference amplifier. Set the FG to 1 kHz and adjust the output to produce a sine wave of ~ 40 mV peak-to-peak. Measure the peak-to-peak output voltage from v(01) to common. Record the ac difference gain and the phase shift between the input and output.

v(in1)        = _____

v(01)         = _____

Ad            = _____

phase shift   = _____

Now reverse the conncetions to v(in1) and v(in2) by connecting the signal to v(in2) and connecting v(in1) to common.

Ad            = _____

phase shift = _____

Question 14.  Explain why there is a difference in phase shift for the different input connections. Which input is the inverting input? Which is the non-inverting input? Is there a difference in gains measured with the signal into different inputs.

Next the ac common mode gain will be obtained. Connect the same FG signal into both inputs. Measure the output, v(01), vs. common on the scope. Obtain the ac common mode gain and the CMRR.

v(cm)         = _____

v(01)         = _____

A(cm)         = _____

Disconnect the difference amplifier.

Question 15.  Explain why the ac and dc CMRR values are different.

## 7-9 OP AMP CHARACTERISTICS

Objectives:   To obtain the CMRR and output resistance of several operational amplifiers.

In unit 5 the offset voltage and input bias current of several op amps were obtained.  The CMRR and output resistance of the same op amps types are measured in this experiment.

For one TL082 op amp measure its offset voltage for use in determining the CMRR.

TL082 V(os) = _____

Connect the TL082 op amp as a voltage follower as shown in figure 7-16. Use the DMM to measure in succession v(in) and (vo-v(in)).  For the latter measurement the DMM is used in the differential mode on the highest sensitivity.  Obtain vo-v(in) values for the nominal v(in) values listed in the table.

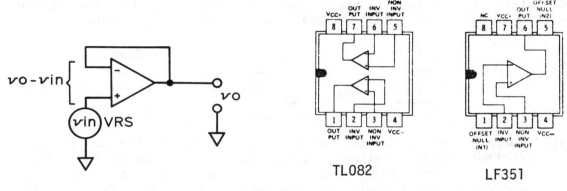

TL082          LF351

Fig. 7-16

| v(in) nominal | v(in) measured | vO-v(in) |
|---|---|---|
| 5 V | _____ | _____ |
| 10 V | _____ | _____ |

Repeat the above measurements for a balanced 351 and a 741 op amp.

| | v(in) nominal | v(in) measured | vo-v(in) |
|---|---|---|---|
| 741 | 5 V | _____ | _____ |
| | 10 V | _____ | _____ |
| 351 | 5 V | _____ | _____ |
| | 10 V | _____ | _____ |

120

To obtain the CMRR divide v(in) by vo-v(in)-V(os).  To obtain the CMRR in dB, use

$$CMRR, dB = 20 \log \frac{v(in)}{vo-v(in)-V(os)}$$

Report your results below.

|  | TL082 | LF351 | µA741 |
|---|---|---|---|
| CMRR, dB | _____ | _____ | _____ |

To measure the follower output resistance, set the VRS at ~ 1 V and connect it as v(in).  Measure the unloaded output voltage vo with the DVM.

Connect a 56 Ω, 5% resistor (measure the exact resistance) as a load RL between the output and common, and measure the output voltage under this load, v(OL).  Calculate the output current from io = v(OL)/RL and the output resistance from Ro = (vo-v(OL))/io.  Repeat for all the op amp types.

|  | TL082 | LF351 | µA741 |
|---|---|---|---|
| vo | _____ | _____ | _____ |
| v(OL) | _____ | _____ | _____ |
| io | _____ | _____ | _____ |
| Ro | _____ | _____ | _____ |

Question 16.  Compare and contrast the three op amps studied as to bias current, offset voltage, CMRR, output resistance and price.  Which op amp would make the best integrator for low level currents?  Which would make the best follower for driving a 100Ω load.  (The 741 is the cheapest per op amp, followed by the TL082 and then the LF351).

# UNIT 8. LINEAR AND NON-LINEAR OP AMP APPLICATIONS

The versatility of the op amp and its more sophisticated relatives is demonstrated in this set of experiments. Difference and instrumentation amplifiers introduce the unit and provide insight into precision amplification. Non-linear elements (diodes) are introduced into common op amp circuit configurations to produce waveshaping circuits and a logarithmic amplifier. The analog multiplier is used to perform analog voltage multiplication, squaring, division, and square root operations. The frequency response of active filters, tuned amplifiers, and oscillators are developed using relatively simple circuits.

The wide range of circuits chosen for study represents a good cross section of the operations achievable with relatively few parts using modern components based upon the op amp.

| Equipment | Parts |
|---|---|
| 1. Basic station | 1. Resistors: 1%, 10 kΩ(4), 100 kΩ(4), 1 MΩ(2); 5%, 1 kΩ, 5 kΩ, 1 MΩ(2) |
| 2. Op amp job board | 2. Capacitors 47 pF(2), 470 pF(2), 0.0047 μF(2), 1 μF |
| 3. Advanced analog circuits job board | 3. 4 signal diodes |
| 4. Utility job board | 4. #349 or #1869 lamp and socket |
| | 5. Photodiode or phototransistor |
| | 6. Potentiometer, 10 kΩ(2) |

## 8-1 DIFFERENCE AND INSTRUMENTATION AMPLIFIERS

Objectives: To study differential input amplifier circuits by wiring op amp difference and instrumentation amplifiers and observing their response to differential and common mode signals.

Wire the difference amplifier shown in figure 8-1. Use the 351 op amp on the op amp job board with the offset balance connected as shown. Use precision resistors on the op amp job board.

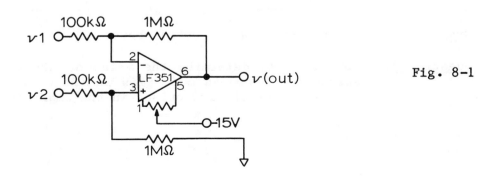

Fig. 8-1

121

Balance the op amp by connecting both v1 and v2 to common and adjusting the offset potentiometer until v(out) is zero.

Leaving v2 connected to common, record v1 and v(out) for a number of values of v1 between +1 and −1 V.

| v1 | v(out) |
|----|--------|
| ——— | ——— |
| ——— | ——— |
| ——— | ——— |
| ——— | ——— |
| ——— | ——— |
| ——— | ——— |
| ——— | ——— |

Question 1. Calculate the average gain of the amplifier. In this measurement, which components are determining the gain of the amplifier? How near is this gain to the value predicted from nominal component values?

Connect v2 to a constant +1 V source. Again measure v(out) for several values of v1 between +1 and −1 V.

| v1 | v2 | v(out) |
|----|----|--------|
| ——— | ——— | ——— |
| ——— | ——— | ——— |
| ——— | ——— | ——— |
| ——— | ——— | ——— |
| ——— | ——— | ——— |
| ——— | ——— | ——— |

Question 2. Repeat question 1 for these data.

Connect both v1 and v2 to the same variable voltage source. Measure v(out) for five or more values of input voltage between +1 and −1 V.

| v1 = v2 | v(out) |
|---------|--------|
| ——— | ——— |
| ——— | ——— |
| ——— | ——— |
| ——— | ——— |
| ——— | ——— |
| ——— | ——— |
| ——— | ——— |

Graph 1. Plot v(out) vs v1 as measured above.

Question 3. From Graph 1, determine the value of the common mode gain, the dependence of common mode gain on the common mode signal level, and the amount of zero offset in the amplifier.

Question 4.    If  all  the  common  mode  gain  were  due  to  imbalance  in  the
               resistance ratios of the two gain determining pairs of resistors,
               determine which pair (inverting or non-inverting) has the higher
               gain, and by what amount.  How could this common mode gain be
               reduced?

Question 5.    Use  the  data  collected  to  calculate  the  common  mode  rejection
               ratio  for  your  difference  amplifier.   Calculate  the  maximum
               common mode signal the amplifier can accept if a 100 mV signal is
               to be amplified with an error of less than 0.1%.

     Wire  the  instrumentation  amplifier  shown  in  figure  8-2  by  adding  the
input voltage followers to the difference amp already connected.  Three 10 kΩ
precision resistors will be needed.

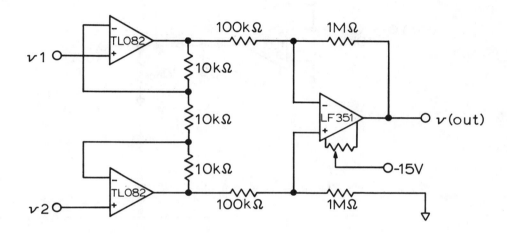

Fig. 8-2

     Check  the  offset  of  the  instrumentation  amplifier  and  adjust  the
difference  amplifier  offset  potentiometer  if  needed.   Measure  v(out)  for
various values of v1 and v2 so that you will be able to determine the
difference gain and the common mode rejection ratio of the instrumentation
amplifier.  Record your data below.  Be sure you have taken all the data
necessary for your calculations.

| v1 | v2 | v(out) |
| --- | --- | --- |
| ____ | ____ | ____ |
| ____ | ____ | ____ |
| ____ | ____ | ____ |
| ____ | ____ | ____ |
| ____ | ____ | ____ |
| ____ | ____ | ____ |

Question 6.    Describe  the  reasoning  you  used  in  selecting  the  values  for  v1
               and v2.  From these data, determine the gain and CMRR.  Explain
               your interpretation of the data.  Compare your results with the
               expected values.

8-2  WAVESHAPING

Objectives:  To introduce the concept of diode waveshaping within the feedback
circuit of op amp amplifiers by wiring limiter and absolute value
circuits and observing the precision waveforms obtained.

Wire the inverting limiter circuit in figure 8-3 on the utility job
board.  Use signal diodes for D1 and D2.  Use a 15 V p-p, zero-centered, 1 kHz
triangular wave for v(in).  The voltage v(R) will be varied between +10 and
-10 V.  Observe v(in) and v(out) on the dual trace oscilloscope as v(R) is
varied.  Sketch the waveform for v(R) values of -4, 0, and +4 V.

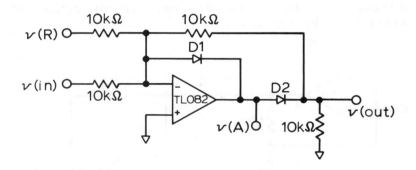

Fig. 8-3

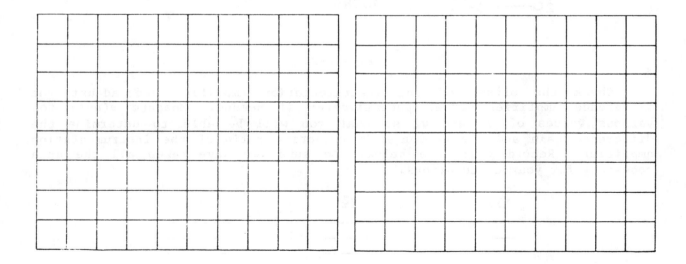

Switch the oscilloscope to X-Y mode to observe the transfer function
(v(out) vs v(in)) for the limiter.  v(out) should be connected to the vertical
channel.  Sketch the transfer function for v(R) = 4 V.  Also observe and

sketch the op amp output v(A) vs v(in) by connecting the vertical channel to v(A).

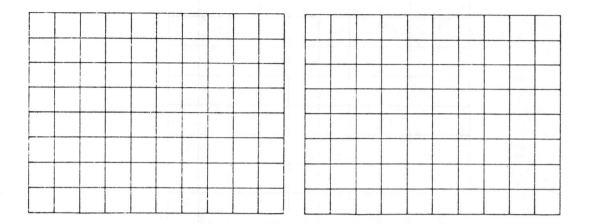

Question 7.  Describe the effect of the value of v(R) on the limiting breakpoint in the waveform, and on the peak positive and negative output signal voltages.

Question 8.  Explain why the forward voltage drop across the conducting diodes does not appear at v(out).

Wire the precision absolute value circuit of figure 8-4 by adding a non-inverting limiter circuit to the inverting limiter previously connected. Note that the v(R) input to the inverting limiter should be removed.

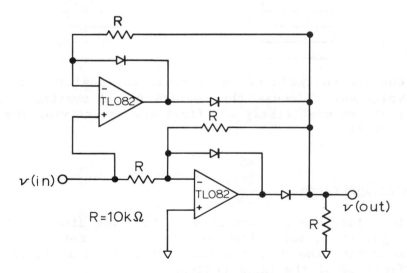

Fig. 8-4

Observe and sketch the transfer function of this circuit.  Use the X-Y mode of the oscilloscope and be sure both scope inputs are in dc mode.

Sketch of transfer function:

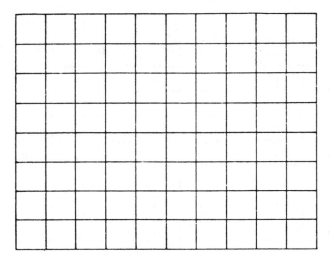

    This absolute value circuit is capable of operating at very low signal levels. Reduce the function generator output voltage (using a resistive divider if needed) while observing the transfer function on the oscilloscope to confirm the circuit's operation at low signal levels.

    Connect v(in) to the ± 1 V VRS output instead of the FG and record the output developed for constant voltage inputs between +1 and −1 V. Be sure to select a few values near zero.

|        v(in)        |        v(out)        |
| ------------------- | -------------------- |
| _____ | _____ |
| _____ | _____ |
| _____ | _____ |
| _____ | _____ |
| _____ | _____ |
| _____ | _____ |

<u>Question 9</u>. Compare the gain of the circuit for positive and negative input voltages. Discuss the amplifier and component characteristics that are most likely to affect the accuracy of the absolute value circuit.

## 8-3  LOGARITHMIC AMPLIFIER

Objectives: To illustrate the use of a pn junction diode feedback element in logarithmic amplifiers and to confirm the Shockley equation by measuring the output voltage of a diode feedback amplifier as a function of the input current.

    Carefully balance the 351 op amp on the op amp job board. Wire the basic logarithmic amplifier (log amp) of figure 8-5.

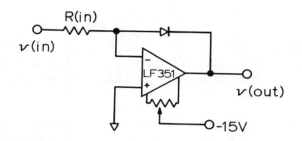

Fig. 8-5

Measure v(out) for the values of v(in) and R(in) shown below.

| v(in) | R(in) | i(in) | v(out) |
|-------|-------|-------|--------|
| 10.0 mV | 1 MΩ | ___ | ___ |
| 10.0 mV | 100 kΩ | ___ | ___ |
| 100 mV | 100 kΩ | ___ | ___ |
| 1.00 V | 100 kΩ | ___ | ___ |
| 10.0 V | 100 kΩ | ___ | ___ |
| 10.0 V | 10 kΩ | ___ | ___ |

Graph 2. Plot v(out) vs log i(in).

Question 10. Discuss the limits to the useful dynamic range of the log amplifier.

Question 11. Compare the slope of the plot in Graph 2 with that predicted by the Shockley equation, $v = (0.059\eta\, T/300)\,(\log i - \log I_0)$ where T is the temperature in K and $\eta$ and $I_0$ are constants.

Connect a second diode in parallel with the original diode, but with the cathode connected to the summing point. This results in a "bipolar" log amplifier -- one which increasingly attenuates signals as their absolute value increases. Use the function generator and the X-Y display of the oscilloscope to determine the response function of this circuit.

Describe the FG and oscilloscope connections and settings for your experiment. Sketch the resulting transfer function below.

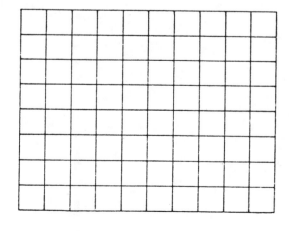

8-4   ANALOG MULTIPLIER

Objectives:   To use an analog multiplier module for multiplication, division,
              and squaring operations to illustrate its transfer functions in
              these modes by observing the multiplier output for several X, Y
              and Z input signals.

     For this experiment and several others, an internally trimmed analog
multiplier such as the AD534 (Analog Devices, Inc., Norwood, Mass.) is most
convenient to avoid the use of several external trimming potentiometers.
     Locate the multiplier on the advanced analog circuits job board as shown
in figure 8-6.   The pinout of the AD534 multiplier is shown in figure 8-7.

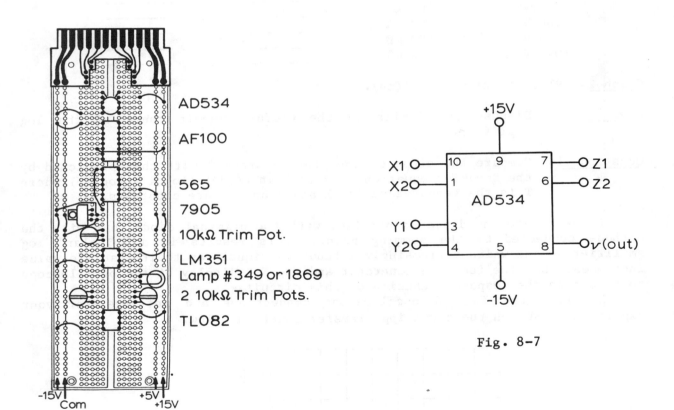

AD534

AF100

565

7905

10kΩ Trim Pot.

LM351
Lamp #349 or 1869
2 10kΩ Trim Pots.

TL082

-15V
Com          +5V
             +15V

Fig. 8-6

Fig. 8-7

For multiplication, use the +10 V output of the VRS as the X1 input and use several fixed voltages from the reference job board as the Y1 input (+10 V, -10 V, -1 V, +1 V). Connect the Z1 input to the output. Connect the X2, Y2 and Z2 inputs to common. Test the multiplier in all four quadrants by applying voltages of both polarities in the range of +10 V. The multiplier transfer function should be v(out) = X1Y1/10. Include in your data set X, Y values of (+10, 0), (0, 0), and (0, +10).

| X1, V | Y1, V | v(out) | Expected v(out) | % Error |
|-------|-------|--------|-----------------|---------|
|       |       |        |                 |         |
|       |       |        |                 |         |
|       |       |        |                 |         |
|       |       |        |                 |         |
|       |       |        |                 |         |
|       |       |        |                 |         |

Question 12. If offsets are included, the equation for the multiplier becomes v(out) = -0.1(vx-v(ox))(vy-v(oy)) + v(oo). Here v(ox) is the X offset, v(oy) is the Y offset, and v(oo) is the output offset. Use your data to evaluate each of these offsets. Explain how the magnitude of offset induced errors changes with X and Y input levels.

To obtain an output voltage proportional to the square of an input voltage, connect the X1 and Y1 inputs to the same voltage source and the X2 and Y2 inputs to common. The Z1 input remains connected to the output. Test the circuit over a +10 V range of VRS voltages and compare to the expected v(out) = v(in)$^2$/10.

| v(in) | v(out) | Expected v(out) | % Error |
|-------|--------|-----------------|---------|
|       |        |                 |         |
|       |        |                 |         |
|       |        |                 |         |
|       |        |                 |         |
|       |        |                 |         |

The "squared voltage" output can be plotted against the input with the X-Y mode of the oscilloscope. Substitute the sine wave output of the FG for the VRS as the source in the squaring circuit wired above. Connect the multiplier output to the vertical scope input and the FG output to the horizontal. Use a 10 Hz sine wave signal. Sketch the resulting display.

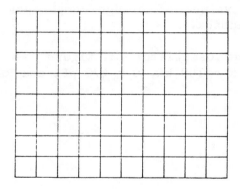

Now use the dual-trace mode to observe the waveforms of the input and output signals. Sketch a representative display and indicate the position of 0 V for each waveform.

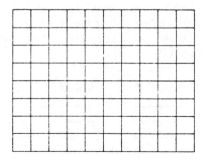

<u>Question 13</u>. Explain the relationship of the frequencies and the dc components of the input and output waveforms.

The division function is achieved by connecting the multiplier output to the Y2 input. (the Z1 input is no longer connected to the output and Z2 is no longer connected to common.) The output voltage will then be $10(Z2-Z1)/(X1-X2) + Y2$. Measure v(out) for several values of (Z2-Z1) and (X1-X2). Grounding Z1, X2, and Y1 is suggested for simplicity. Errors will be generally worse for small values of (X1-X2).

| v(Z2)-v(Z1) | v(X1)-v(X2) | v(out) |
| --- | --- | --- |
| _____ | _____ | _____ |
| _____ | _____ | _____ |
| _____ | _____ | _____ |
| _____ | _____ | _____ |
| _____ | _____ | _____ |
| _____ | _____ | _____ |

<u>Question 14</u>. The output limits of the AD534 are $\pm$11 V. Calculate and plot the minimum value for V(X1)-V(X2) as a function of V(Z2)-V(Z1) over the V(Z) range of $\pm$10 V.

## 8-5 WIEN BRIDGE OSCILLATOR

Objectives: To illustrate the principle of regenerative oscillation by wiring a Wien bridge oscillator and observing the frequency and waveform for various values of bridge components and various amounts of negative feedback.

Wire the Wien bridge oscillator of figure 8-8 on the advanced analog circuits job board. Use R = 100 k$\Omega$, C = 470 pF, and RA = 10 k$\Omega$ potentiometer.

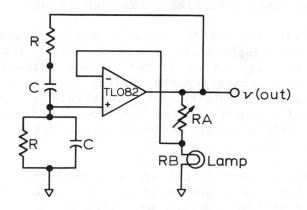

Fig. 8-8

Observe the output on the oscilloscope. Adjust RA until a sine wave is obtained. Measure the frequency and the peak-to-peak voltage of the oscillator output for the values of R and C listed below. Hint: When RA is properly adjusted, the lamp is off. Once it is adjusted properly, leave it for the remainder of this experiment.

| R | C | Measured f(out) | Calc. f(out) | v (p-p) |
|---|---|---|---|---|
| 100 kΩ | 470 pF | | | |
| 10 kΩ | 470 pF | | | |
| 10 kΩ | 0.0047 μF | | | |
| 100 kΩ | 0.0047 μF | | | |
| 100 kΩ | 47 pF | | | |

Question 15. Compare the measured values of f(out) with the expected value of f(out) = $1/(2\pi RC)$, and account for any discrepancies.

Question 16. What is the purpose of the lamp in this circuit?

Replace the lamp with a 10 kΩ potentiometer (RB). Adjust its resistance until oscillation begins. Very carefully adjust RB until a sine wave with low distortion is obtained. Note any remaining distortion observable with the scope. If a spectrum analyzer is available, obtain the frequency spectrum of the oscillator output with the potentiometer for RB and with the lamp for RB.

Question 17. Comment on the quality of the output obtained with the lamp and with the 10 kΩ potentiometer for RB. Why is the value of the RA/RB ratio so critical for a high quality sine wave?

## 8-6 ACTIVE FILTERS, TUNED AMPLIFIERS, AND OSCILLATORS

Objectives: To investigate state variable active filters by measuring the frequency response of the high pass, low pass and band-bass outputs of an active filter IC, to investigate tuned amplifier circuits by adjusting the quality factor of the filter to achieve a resonance frequency, to investigate a notch filter by summing high and low pass filter outputs in a summing amplifier, and to investigate a state variable oscillator by providing a positive feedback loop.

The AF100 universal active filter is a versatile state variable active filter. It has high pass, low pass and bandpass outputs simultaneously available and an uncommitted summing amplifier for making notch filters. The center frequency is tunable from 200 Hz to 10 kHz with 2 resistors. The quality factor (Q) is variable from 0.01 to 500 by changing two additional resistors. The AF100 can be used in either an inverting or a non-inverting configuration.

Locate the AF100 on the advanced analog circuits job board. The pinout is shown in Figure 8-9 along with a circuit diagram. The center frequency fo of the filter is programmed by identical resistors Rf1 and Rf2 according to the equation fo = $(50.33 \times 10^6)/Rf)$, Hz.

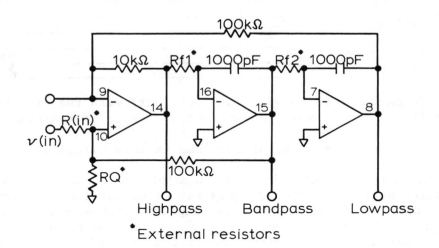

Fig. 8-9

The gain and Q value are determined by resistors R(in) and RQ. All components except these 4 resistors are internal.

Wire the non-inverting mode filter circuit of figure 8-9 using external precision resistor values of Rf1 = Rf2 = 100 kΩ and R(in) = RQ = 100 kΩ. Details of the external connections are shown in figure 8-10.

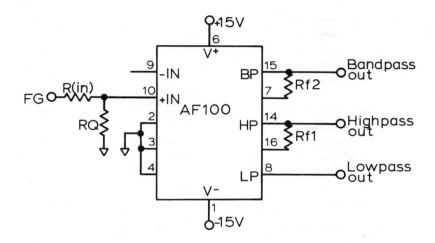

Fig. 8-10

The external resistor values should give a center frequency of ≃ 500 Hz and a Q of slightly greater than unity.  Connect the FG to R(in) and use the scope to observe the FG output and the bandpass output of the filter.  Connect the FG TTL output to the CTFM for a digital readout of frequency.  Set the FG for a 1 V peak-to-peak sine wave.  Observe the bandpass output as the FG frequency is varied through the center frequency, fo.

What happens to the bandpass output at fo? _____

To measure fo accurately, switch the scope to produce an X-Y plot (Lissajous figure) of filter input vs. filter output.  At the center frequency the bandpass output should be exactly $180^\circ$ out of phase with the input signal. Use the Lissajous figure to adjust the FG exactly to the center frequency (refer to page 43 of the text for Lissajous figures).  Read the exact center frequency from the CTFM.

fo _____

Now switch the scope back to the dual trace mode and measure the peak-to-peak output voltage of the bandpass filter as a function of FG frequency over a range of 20 Hz to 20 kHz.  Record 10-15 values in this range including several near fo.  Calculate the filter gain, v(out)/v(in) in dB.

134

| FG freq. | v(in) | v(out) | gain, dB |
|----------|-------|--------|----------|
| ———— | ——— | ——— | ——— |
| ———— | ——— | ——— | ——— |
| ———— | ——— | ——— | ——— |
| ———— | ——— | ——— | ——— |
| ———— | ——— | ——— | ——— |
| ———— | ——— | ——— | ——— |
| ———— | ——— | ——— | ——— |
| ———— | ——— | ——— | ——— |
| ———— | ——— | ——— | ——— |
| ———— | ——— | ——— | ——— |
| ———— | ——— | ——— | ——— |
| ———— | ——— | ——— | ——— |
| ———— | ——— | ——— | ——— |
| ———— | ——— | ——— | ——— |
| ———— | ——— | ——— | ——— |

Graph 3.  Plot the filter gain in dB vs. log frequency.

Question 18.  From the slopes of the plot in graph 3, away from fo, determine the rolloff rate of the filter in dB/decade on both sides of fo.

Now connect the scope to the low pass filter output.  Measure the filter gain and phase shift as a function of frequency in the range of 20 Hz to 20 kHz.  Use the Lissajous method to determine phase shift (see page 44 of the text).  Accurately determine the 3dB frequency (frequency where gain = 0.707 low frequency gain).

| Freq. | v(in) | v(out) | b | c | $\theta$ | Gain, dB |
|-------|-------|--------|---|---|----------|----------|
| ——— | ——— | ——— | ——— | ——— | ——— | ——— |
| ——— | ——— | ——— | ——— | ——— | ——— | ——— |
| ——— | ——— | ——— | ——— | ——— | ——— | ——— |
| ——— | ——— | ——— | ——— | ——— | ——— | ——— |
| ——— | ——— | ——— | ——— | ——— | ——— | ——— |
| ——— | ——— | ——— | ——— | ——— | ——— | ——— |
| ——— | ——— | ——— | ——— | ——— | ——— | ——— |
| ——— | ——— | ——— | ——— | ——— | ——— | ——— |
| ——— | ——— | ——— | ——— | ——— | ——— | ——— |
| ——— | ——— | ——— | ——— | ——— | ——— | ——— |
| ——— | ——— | ——— | ——— | ——— | ——— | ——— |
| ——— | ——— | ——— | ——— | ——— | ——— | ——— |
| ——— | ——— | ——— | ——— | ——— | ——— | ——— |
| ——— | ——— | ——— | ——— | ——— | ——— | ——— |

f(3dB)  _____

Graph 4.  Plot the gain and phase shift vs. log frequency.

Question 19.  What is the phase shift at the 3dB frequency?  Determine the order n of the filter from the rolloff slope (slope in dB/decade = 20 n).

Now observe the high pass output and qualitatively determine that it behaves as a high pass filter. Accurately determine the 3dB point and the phase shift of the high pass output at the 3dB frequency.

f(3dB) _____

θ at f(3dB) _____

Next the values of R(in) and RQ will be changed to make a filter of much higher Q. Set the FG to give a p-p sine wave of ≃ 0.5 V. Change R(in) to 20 kΩ and RQ to 1 kΩ. Observe the bandpass output and the low pass outputs simultaneously on the scope. If the low pass output is clipped, reduce the FG amplitude until no distortion is obtained. Measure v(out) for both outputs vs. frequency as before.

| freq. | v(in) | Bandpass | | Lowpass | |
|-------|-------|----------|---------|---------|---------|
| | | v(out) | Gain, dB | v(out) | Gain, dB |
| _____ | _____ | _____ | _____ | _____ | _____ |
| _____ | _____ | _____ | _____ | _____ | _____ |
| _____ | _____ | _____ | _____ | _____ | _____ |
| _____ | _____ | _____ | _____ | _____ | _____ |
| _____ | _____ | _____ | _____ | _____ | _____ |
| _____ | _____ | _____ | _____ | _____ | _____ |
| _____ | _____ | _____ | _____ | _____ | _____ |
| _____ | _____ | _____ | _____ | _____ | _____ |
| _____ | _____ | _____ | _____ | _____ | _____ |

Graph 5. Plot the gain in dB vs. log frequency for both outputs.

Question 20. Estimate the Q of this filter and that of the previous bandpass filter. Q can be measured as the ratio of the center frequency fo of the bandpass output to the bandwidth (width in Hz from lower 3dB point to upper 3dB point). Why does the low pass frequency response have a peak in it?

Return the AF100 to the low Q state (R(in) = 100 kΩ, RQ = 100 kΩ). Change the feedback resistors to the values shown below and measure the center frequency of the bandpass output.

| Rf1 = Rf2 | Center freq. |
|-----------|--------------|
| 10 kΩ | _____ |
| 50 kΩ | _____ |
| 200 kΩ | _____ |

Next a notch filter will be constructed by summing the low and high pass outputs with the uncommitted summing amplifier. Set the AF100 to produce a center frequency of 500 Hz. Sum the outputs as shown in figure 8-11.

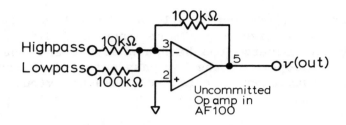

Fig. 8-11

(Be certain to remove the common connection to pin 3). Measure the frequency response of the notch filter for FG frequencies in the range of 100 Hz to 1 kHz.

| freq. | v(in) | v(out) |
|-------|-------|--------|
| _____ | _____ | _____ |
| _____ | _____ | _____ |
| _____ | _____ | _____ |
| _____ | _____ | _____ |
| _____ | _____ | _____ |
| _____ | _____ | _____ |

Question 21. Describe a use of the notch filter in reducing noise. What types of noise can be reduced?

Disconnect the summing amplifier of figure 8-11. Reconnect pin 3 of the AF100 to common. Connect an inverting mode filter by moving R(in) and the FG from the non-inverting input (see figure 8-9) to the inverting input (pin 9). Leave RQ connected between pin 10 and common. Qualitatively show that the filter functions in a similar manner to the non-inverting mode. Note carefully any differences in phase at the outputs compared to the non-inverting mode filter. To make a very high Q filter, change R(in) to 20 kΩ and RQ to 100 Ω.

Phase difference _____

Return the filter to low Q state, but leave it in the inverting mode. Disconnect the FG.

Question 22. When might the inverting mode be used instead of the non-inverting mode?

The inverting mode filter can be readily converted to a state variable oscillator, by providing a positive feedback loop and some gain adjustment. Wire the circuit of figure 8-12 using the uncommitted summing amplifier for the follower-with-gain. The 10 kΩ potentiometer provides the gain adjust for the loop, while the bandpass filter (set to a center frequency of ≃ 500 Hz) provides frequency selectivity. Connect the output to the scope. Begin with the potentiometer set to provide nearly the full 10 kΩ resistance. Slowly

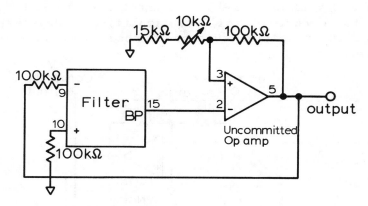

Fig. 8-12

turn the pot towards lower resistance values (higher gains) until oscillation is observed on the scope. Too much gain will produce distortion of the sine wave output. If this is observed, very slowly increase the resistance of the pot until a distortion-free sine wave is observed. This may require several back and forth adjustments to set the appropriate gain. Report the oscillation frequency from the scope measurement.

f(osc) _____

Question 23. Describe what modifications could be made to the amplifier so that the gain would be automatically adjusted. How can the oscillation frequency be changed?

Question 24. How could the non-inverting mode filter be made into an oscillator? Give a circuit diagram. What are the conditions for oscillation? Could a high Q filter be made to oscillate? How? What would the gain of the amplifier have to be for the circuit to oscillate with a high Q filter? How could this be implemented in the non-inverting filter mode? In the inverting filter mode?

138

## 8-7 LOCK-IN AMPLIFIER

Objectives: To demonstrate how modulation, band-pass limitation, and phase-locked demodulation can be used to discriminate against noise components in a signal by constructing a lock-in amplifier with a tuned amplifier and multiplier and using it to observe the component of a photodetector output that results from a modulated LED.

Arrange in left-to-right order in the frame the BSI job board, the advanced analog gates job board, and the REF job board. Referring to the circuit diagram of figure 8-13, connect a biased photodiode (or phototransistor) to a current follower amplifier. Physically arrange the photodiode so that its sensitive surface is about 1 cm from one of the LED's on the BSI job board.

Fig. 8-13

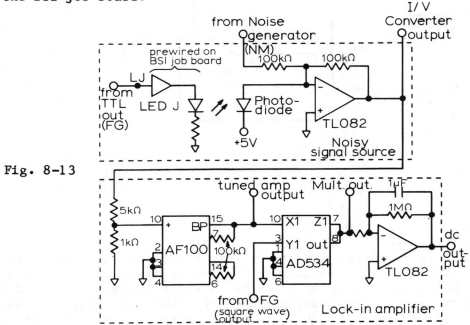

Connect the input of the LED to the TTL output of the FG. Set the FG frequency to about 500 Hz. Do not connect the noise generator to the current follower yet, but observe the I-to-V converter output, both dc and ac components, with the oscilloscope. Adjust the position of the photodiode (or the current follower gain, if necessary) to obtain a square wave component of the I-to-V converter of 50 mV to 100 mV.

dc level _____ V

square wave amplitude _____ V (p-p)

approximate noise level _____ V (p-p)

Question 25. Why is there a dc component in the current-to-voltage converter output?

Now connect the AF100 tuned amplifier circuit to the output of the current follower. (Do not connect the noise generator yet). Observe the tuned amplifier output with the scope. Adjust the FG frequency to obtain a maximum output from the tuned amplifier. Record this value. What is the dc component?

tuned amp output     _____ V (p-p)

Waveshape?     _____

dc component     _____ V

FG frequency setting _____ Hz.

<u>Question 26</u>. You should have observed a sine wave at the tuned amplifier output. The input, however, was a noisy square wave with a dc component. Explain the difference in input and output waveforms.

The analog multiplier and low pass filter (phase-locked demodulator) should now be connected. The tuned amplifier output is multiplied by a square wave that is synchronous with the LED modulation. Adjust the FG square wave output to supply a $\pm$ 10 V reference signal to the multiplier. Observe the multiplier output. Adjust the FG frequency carefully to obtain a waveform that most closely approximates a full-wave rectified sine wave. Draw the observed multiplier output waveform. Label the axes.

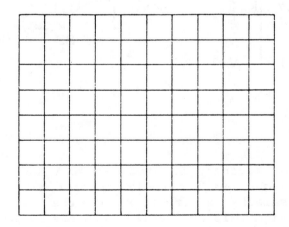

Connect the active low-pass filter to the multiplier output. Observe the dc output with the DMM. Record the dc level observed with the light path between LED and detector clear and then obstructed.

clear path output     _____ V

obstructed path output _____ V

Look at the I-to-V output again and observe the ratio of the square wave amplitude to the noise amplitude. Observe also the range of values displayed for the dc output by the DMM over a few minutes. Compare the signal-to-noise ratios of the amplified signal and the lock-in amplifier output.

I-to-V output square wave/I-to-V output noise _____

dc output voltage/dc output variation _____

<u>Question 27</u>. Calculate the signal-to-noise ratio improvement obtained with the lock-in amplifier.

To better demonstrate the noise rejection capabilities of the lock-in amplifier, still more noise will be intentionally added to the signal. This noise will be obtained from the noise generator circuit on the REF job board. This circuit will be introduced before it is applied to the lock-in amplifier. The circuit of the random noise generator on the reference job board is shown in figure 8-14.

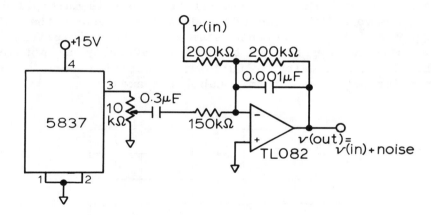

Fig. 8-14

The 5837 digital noise generator IC produces 10 V pulses that have varying durations. The pulse durations are random integer multiples of 20 μs. The 10 kΩ potentiometer selects a fraction of the noise generator output amplitude. The noise signal is ac coupled into a summing amplifier that also serves as an active low-pass filter. The additional input to the summing amplifier allows the noise generator signal to be added to another signal to produce signals with a variable amount of noise.

Vary the 10 kΩ potentiometer to obtain maximum noise amplitude. Sketch the waveform observed on both sides of the coupling capacitor and at the output of the summing amplifier for maximum noise amplitude. An oscilloscope time base of 20 μs/div is recommended.

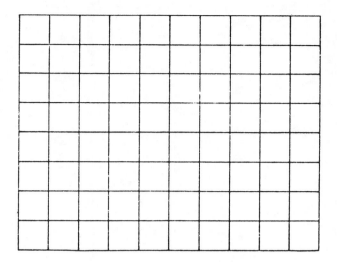

Also observe the output of the summing amplifier at a sweep speed of 500 μs/div. This output will be called the noise mixer (NM) output.

Question 28. Calculate the cut-off frequency of the low-pass filter in the NM. What is the attenuation of this filter for the frequency component that results from transitions every 20 μs (25 kHz)?

Connect the NM output through a 100 kΩ resistor to the current follower summing point of the noisy signal source circuit shown in figure 8-13. Observe the current follower output with a scope and adjust the noise generator output from zero to a value at which the square wave becomes difficult to distinguish. (The scope should be triggered from the square-wave or TTL output of the FG to avoid loss of synch.) Measure the dc output voltage of the lock-in amplifier and observe its variation for a few minutes. Estimate a peak-to-peak value for the noise and compare again the signal-to-noise ratios of the input and output of the lock-in amplifier.

I-to-V output square wave/I-to-V output noise _____

dc output voltage/dc output variation _____

Question 29. Explain why it is necessary to modulate the signal in order to obtain the improvement in signal-to-noise ratio through the lock-in technique.

# UNIT 9. FREQUENCY, TIME, AND THE INTEGRATING DVM

The experiments in this unit involve operations with time domain signals. They begin with an intensive study of the conversion of frequency and period domain signals to the digital domain through the counter-timer-frequency meter (CTFM or universal counter). The Schmitt trigger is introduced as a comparator with hysteresis to avoid false triggering through noise. A charge-to-count conversion is shown to be the basis of the voltage-to-frequency converter (VFC), a conversion into one of the time domains. The use of the time domains in analog-to-digital conversion is illustrated through the combination of the VFC and the digital frequency meter and through the dual-slope converter in a digital panel meter. The unit concludes with the phase-locked loop -- a servo system that can perform precision multiplication of a frequency domain signal by a number encoded in the digital domain.

| <u>Equipment</u> | <u>Parts</u> |
|---|---|
| 1. Basic station | 1. Optointerrupter |
| 2. Digital voltmeter (DVM) | 2. Resistors: 150 Ω, 220 Ω, 1 kΩ, 15 kΩ, 47 kΩ |
| 3. Reference job board | |
| 4. BSI job board | 3. Capacitors: 0.01 µF |
| 5. Basic gates job board | 4. Punched IBM card |
| 6. Op amp job board | 5. Loose IC (can be dead) |
| 7. Signal conditioning job board | 6. Unknown capacitor |

## 9-1 COUNTING, TIMING AND FREQUENCY RATIOING

Objectives: To study in detail the functions and operation of a counter-timer frequency meter and to illustrate the process of exploring the characteristics of a large-scale integrated circuit (LSI) device by applying the CTFM based on the Intersil ICM 7226 Universal Counter System IC and by interpreting the results in terms of the counting measurement process for each mode.

### A. Introduction

A somewhat simplified block diagram of the CTFM is shown in figure 9-1. The versatility of this system arises from the interconnection of its various subunits in a variety of configurations. This results in the five functions shown in the table.

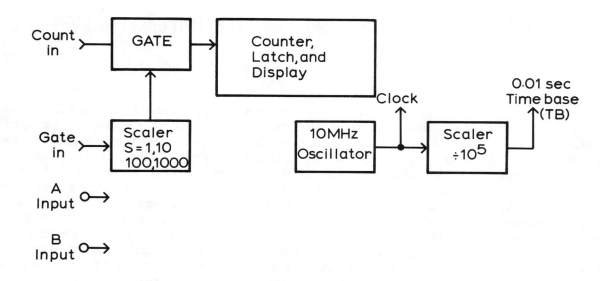

Fig. 9-1

| Function | Count in | Gate in | Total count | Units displayed |
|----------|----------|---------|-------------|-----------------|
| Frequency | A | TB | $0.01 \, F(A)S$ | kHz |
| Period | clock | A | $10^7 \, T(A)S$ | s |
| Ratio A/B | A | B | $F(A)S/F(B)$ | $F(A)/F(B)$ |
| Time int. | clock | $T(A-B)$ | $10^7 \, T(A-B)S$ | s |
| Units | A | hold/reset | $C(A)$ | --- |

In the frequency mode, the A input signal is connected to Count in, and the 0.01 s time base signal (TB) is connected to Gate in. The counter then counts the number of cycles of signal A that occur during the gate interval. The factor S of the gate scaler may be 1, 10, 100, or 1000 as set by the range switch, which produces count intervals of 0.01 s, 0.1 s, 1 s, or 10 s, respectively. The total count displayed is thus the number of cycles of A that occur in 0.01 s (0.01 F(A), where F(A) is the frequency of signal A) times the scaling factor S. The decimal point of the display is automatically set to display the frequency in kHz.

In the period mode, the oscillations of the 10 MHz clock that occur in S cycles of the signal are counted. The CTFM does this by connecting Clock to Count in and signal A to Gate in. The counter accumulates $10^7$ counts for each second of the gate interval T(A)S, where T(A) is the period of the A signal in seconds. The display decimal point is set to read the period in microseconds.

The frequency ratio measurement is achieved by connecting signal A to Count in and signal B to Gate in. Signal B is scaled by S so the count that results is the number of cycles of A that occur in S cycles of B. The display reads in whole numbers when S = 1, in tenths when S = 10, and so on. The time interval function measures the time between a HI-LO transition of signal A and a subsequent HI-LO transition of signal B. The gate interval is thus started by A and ended by B. The counter counts tenths of microseconds from the clock during this interval. The counter will accumulate the counts from S gate intervals to provide more significant figures when needed. The counter is not latched and reset until the S gate intervals have occurred. The measurement of single intervals in ths mode must be preceeded by HI-LO transitions at A and B to "arm" the gate interval triggers.

In the count mode, input A is connected to Count in and the gate is open unless the hold button is depressed. The reset button restores the counter contents to zero and keeps it reset as long as it is depressed. Release of the reset button can be used to begin the count, which can then be halted by the hold button or read out.

B. Automatic Measurement Cycle of the CTFM

On the CTFM, locate the Reset Out, Measure, and Store contacts. Connect the FG TTL output to the A input of the CTFM. Set the FG frequency to 90 kHz and the CTFM in the frequency mode, 0.01 s range. Observe the CTFM meas output in the scope with the scope sweep rate set to 50 ms/div. Trigger on the negative slope of meas. In the table below, record the CTFM display for each position of the CTFM range switch. Also record the duration of the LO pulse at the meas output.

| Range, s | Display | t(meas), s | Calculated gate time, s |
|----------|---------|------------|-------------------------|
| 0.01     | _____ | _____ | _____ |
| 0.1      | _____ | _____ | _____ |
| 1.0      | _____ | _____ | _____ |
| 10.0     | _____ | _____ | _____ |

(use x-y scope mode and your watch)

Question 1. Calculate the measurement gate time (time interval during which the FG pulses are counted for each range setting). Compare with t(meas) the observed gate. Explain any discrepancy.

Switch the CTFM to period mode, and the range switch to 0.01 s. Simultaneously observe meas and the FG TTL output on the scope. Trigger on the falling edge of meas. Sketch the waveforms for these settings.

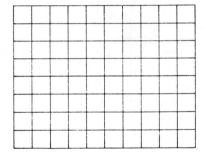
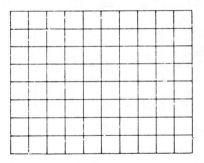

Set the range switch at 0.1 s. Record the number of FG TTL output cycles which occur while meas is LO, and the duration of t(meas).

Number of cycles _____

t(meas)LO _____

Change the FG to 60 kHz. Again the number of cycles and t(meas)

Number of cycles _____

t(meas)LO _____

For each position of the range switch record the value displayed and the duration of t(meas).

| Range, s | Display | t(meas), s | Calculated number of FG TTL cycles/gate period |
|---|---|---|---|
| 0.01 | _____ | _____ | _____ |
| 0.1 | _____ | _____ | _____ |
| 1.0 | _____ | _____ | _____ |
| 10.0 | _____ | _____ | _____ |

<u>Question 2</u>. From display, t(meas) and the TTL frequency, show the calculation which gives the CTFM timebase frequency. Also calculate the number of input periods per gate period at each range setting.

Keep the FG frequency set to 60 kHz, and set the CTFM range switch to 10 s. Disconnect the FG TTL output from the second channel of the scope and connect instead the CTFM store output. This output is LO when the CTFM is latching the count from the just-completed gate period. Trigger on meas, and observe both waveforms on the 500 ms/div. Sketch the waveforms below. Leave room in the sketch for a third waveform below these two.

Move the second scope channel input lead from store to the CTFM reset output. Add the reset out waveform to the sketch above, preserving the time relationship.

<u>Question 3</u>. Describe the complete period measurement sequence. Give the time of each operation in order of its execution to as high a degree of accuracy as you can.

The interval when meas is HI is used by the 7226 IC for housekeeping chores, such as determining the position of range and function switches, latching the count, resetting the counter, and readying the gate for the next measurement. Design and carry out an experiment that verifies that the length of this interval is independent of gate period. Describe your experiment and the results.

## C. Significant Digits in Frequency and Period Measurements.

The precision of a measurement of frequency or period depends upon the frequency of the unknown signal and the gate time of the measurement. Connect the 1 MHz output of the precision time base on the REF job board to the A input of the CTFM. Record the CTFM display for the measurement of the period and the frequency of the 1 MHz signal on all four CTFM ranges specified. Repeat for the other time base frequencies shown in the table.

| | Period | | | | Frequency | | | |
|---|---|---|---|---|---|---|---|---|
| REF freq. | 0.01 s | 0.1 s | 1.0 s | 10 s | 0.01 s | 0.1 s | 1 s | 10 s |
| 1 MHz | | | | | | | | |
| 10 kHz | | | | | | | | |
| 100 Hz | | | | | | | | |
| 1 Hz | | | | | | | | |

148

Question 4. The number of significant digits in a frequency calculated from a period is equal to the number of significant digits in the period. Therefore the most efficient measurement is to choose the period or frequency mode based on which one gives the largest number of significant digits in the least time. Justify the following statement: "For any frequency, it is always more efficient to measure period than frequency if any number of periods may be averaged and if the clock frequency is equal to the maximum counter frequency."

Question 5. In a practical CTFM, the number of periods which may be averaged and the number of time base frequencies available is limited. Therefore it is not always possible to achieve the same number of significant digits in period and frequency modes. This affects the selection of period frequency mode in a practical measurement. In the CTFM used in this experiment:
a) What is the maximum number of significant digits obtained in the measurement of a 1 MHz signal by frequency and by period measurements?
b) Compare the gate time required to achieve 5 significant digits in the measurement of a 1 MHz signal by period and by frequency measurements.
c) Repeat question a) for a 10 Hz signal.
d) Suppose that noise in the comparator input signal produces a random single period time uncertainty (jitter) equal to 1% of the period. What is the maximum number of significant digits that could be obtained in a period measurement of a single cycle? How many periods are required to produce 4 significant digits? How many significant digits are produced in a frequency measurement made for which the gate time is equal to the gate time in multiple period average measurements that produce four significant digits?

D. Frequency Ratio Measurements

Connect CTFM input A to the precision time base output. Set the time base to 1 MHz. Connect CTFM input B to the FG TTL output. Set the CTFM to the frequency ratio mode (A/B). Measure the frequency ratio for the FG frequencies and CTFM ranges in the table below. Indicate the uncertainty in repeated measurements.

| Frequency | Range | | | |
|---|---|---|---|---|
| | 0.01 | 0.1 | 1.0 | 10.0 |
| 900 kHz | | | | |
| 200 Hz | | | | |

Question 6. Calculate the number periods of the signal at the B input at each position of the range switch. Discuss the relative uncertainty observed as a function of number of periods ratioed.

E.  Time Interval Measurements

Connect the 555 astable circuit of figure 9-2.  Use the 555 located on the signal conditioning job board.  Obtain an inverted 555 output signal using an inverter or inverting gate on the basic gates job board.

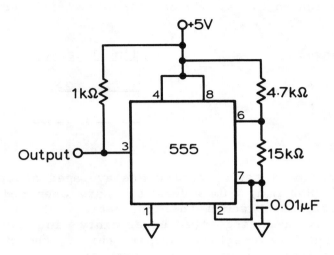

Fig. 9-2

Display the 555 output and its inverse on scope channels 1 and 2, respectively.  Measure from the scope display the HI and LO intervals, and the period of the 555 output.

HI time _____

LO time _____

Period _____

In the CTFM time interval mode, the falling edge of input A starts the measurement, and the falling edge of input B stops it.  The 10 MHz oscillator is counted during the measurement.  Connect the 555 output to input A, and measure the period of the 555 output.  Connect the inverted 555 output to the CTFM B input.  Set the CTFM to time interval mode.  Set the range switch to 0.01.  Record the interval.  What are the units of the readout?  Which portion of the 555 output signal (high or low) is measured?

Period measured on CTFM _____

_____ (HI,LO) portion of 555 output lasts _____

Reverse the CTFM A and B inputs and record the time interval and indicate which portion of the output was measured.

_____ (HI,LO) portion of 555 output lasts _____

Question 7.  Compare the scope period and CTFM period measurements.  Calculate the asymmetry of the 555 output from the scope data and from the CTFM data in terms of the percent of the time the output is HI. How nearly is the period, as measured by the CTFM, equal to the sum of the HI and LO intervals?

Set the range switch to 0.1 s.  Record the measured interval.  Repeat for 1.0 s and 10 s.

| Range | Time Interval |
|-------|---------------|
| 0.1 s | _____ |
| 1.0 s | _____ |
| 10 s | _____ |

Question 8.  How many time interval cycles are averaged at each range switch setting?  How many times does the gate open and close for each range setting?  Discuss the implications of this procedure with respect to reducing the uncertainty in the time interval measurement by averaging.  Compare this to the effect of multiple period averaging in period measurements.

F.  Counting

Units counting is initiated by pressing the CTFM reset button.  Connect the CTFM A input to the precision time base output.  Set the time base to 1 kHz.  Set the CTFM Run/Hold toggle switch to run.  Press and release the reset push button.  Observe the accumulation of counts from the DTB.  Set the Run/Hold switch to hold.  Observe the effect.  Set the Run/Hold switch to run. Observe that the count continues without any reset.  Select an appropriate time base and devise and test a sequence of operations which will allow the CTFM to be used as a stop watch reading in hundreths of seconds.  The Run/Hold switch should be used to start and stop the watch.  Describe your procedure and results below.

## 9-2 SCHMITT TRIGGER

Objectives: To demonstrate the triggering difficulties with comparator circuits when used for counting and timing measurements and to illustrate the characteristics of the Schmitt trigger by observing the limits of comparator triggering reliability and contrasting those with the behavior of a Schmitt trigger wired from the same comparator.

### A. Comparator Limitations

Obtain the op amp job board and identify the LM 311 comparator IC. The normal schematic for the circuit is shown in figure 9-3.

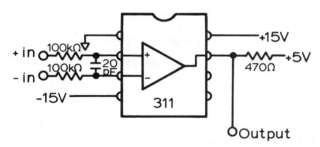

Fig. 9-3

Connect the + input to common and the FG output to the - input. Obtain a 100 kHz 10 V p-p triangular wave output from the FG. Use the TTL output of the FG to trigger the scope. Observe the comparator output and the FG output on the 2 channels of the scope. Adjust the FG offset control so that the comparator output duty cycle is about 50% (duty cycle is the fraction of the cycle time the output is HI). Set the time base at 1 μs/div, set the sensitivity of the channel monitoring the FG to 0.1 V/div, and the comparator output channel to 1 V/div. Set the FG channel zero position at the center line of the display. Turn on the sweep magnifier and adjust the horizontal position to examine the region of the LO-HI transition of the comparator output. What is the delay between the zero-crossing of the FG output and the beginning of the comparator output change? What is the delay time for the comparator output to reach 3 V?

    Delay to beginning _____,
    Delay to 3 V       _____

Repeat for the falling edge.

    Delay to beginning _____,
    Delay to 1 V       _____

Leave the connections to the scope and FG as they are, connect the A input of the frequency meter to the comparator output and the B input to the TTL output of the FG. Measure the frequency ratio (A/B) to the maximum number of significant figures. The ratio should be exactly one if the comparator is being triggered properly.

    Observed frequency ratio _____

152

    Reduce the amplitude of the FG output until the frequency ratio is no
longer exactly one.  Sketch the displays of both traces and indicate the
probable cause of trigger failure.  Several causes are possible.  They include
noise in the signals and oscillation in the comparator.  The latter is a high-
gain, high-speed amplifier that is quite prone to oscillation in the
transition region.

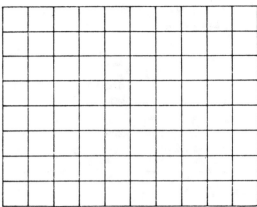

Vary the amplitude of the FG over the range of its amplitude control.  Measure
the amplitude at which the triggering becomes erratic for FG frequencies of
100 kHz, 10 kHz, 1 kHz, 100 Hz.  Calculate the slope of the triangular wave
signal at each frequency.

| signal level | slope, dV/dt |
|---|---|
| _____ V, 100 kHz | _____ V/s |
| _____ V, 10 kHz | _____ V/s |
| _____ V, 1 kHz | _____ V/s |
| _____ V, 100 Hz | _____ V/s |

Perform a brief experiment to indicate the improved reliability of triggering
from a square wave vs. the triangular wave.  Describe your experiment and
explain the results.

    Description and Explanation:

Question 9.   Comment on the relationship between signal slope and triggering
              reliability.  For a sine wave signal, what part of the waveform
              would produce the most reliable triggering?

## B. Schmitt Trigger

One way to reduce the possibility of erratic triggering from noise and oscillation is positive feedback. This introduces hysteresis in the response and produces a circuit called the Schmitt Trigger. Connect the circuit shown in figure 9-4. Here the Schmitt trigger is wired using an op amp, and the LM 311 is used to provide a TTL compatible output (see Unit 4). The Schmitt circuit can be used with the comparator directly in applications, but careful component selection and layouts are required to avoid oscillations with slowly changing input signals. Connect the FG output to v(in) and connect the v(ref) input to common. Leave the frequency meter connected to monitor the FG-to-comparator output frequency ratio.

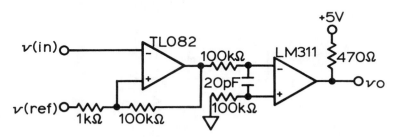

Fig. 9-4

Set the FG to produce a 1.5 V p-p triangular wave at 1 kHz. Adjust the offset so that the comparator output has a 50% duty cycle. From the scope display of the input and output waveforms, measure the FG output voltage levels at which the LO→HI and HI→LO comparator output transitions occur. Also measure the HI and LO levels at the comparator output.

_____ V, LO→HI

_____ V, HI→LO

_____ V, HI.

_____ V, LO

<u>Question 10</u>. Compare the observed hysteresis level with the value expected from the theory. (Remember that the op amp output ($\pm$ 12 V) is providing the feedback).

Adjust the FG amplitude and offset for the minimum amplitude 100 kHz triangular wave that produces reliable triggering as indicated by scope observation and confirmed by frequency ratio measurement. Verify the reliability of the triggering for the same settings of FG amplitude and offset at frequencies of 10 kHz, 1 kHz, 100 Hz, 10 Hz and 1 Hz. Record your results below.

Results:

Leave the Schmitt trigger (TL082 and LM 311 circuit) connected for the next section.

C.  Schmitt Trigger Application

Connect the optointerrupter circuit of figure 9-5 and patch wire its output to the v(in) input of the Schmitt trigger of figure 9-4.  Connect the $\pm$ 10 V VRS output to the v(ref) input.

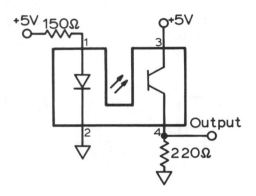

Fig. 9-5

Display the output of the optointerrupter (Ch 1) and the output of the comparator (Ch 2) on the scope.  With nothing interrupting the light beam, adjust the VRS until the comparator output just goes LO.  Pass a wire back and forth through the interrupter and adjust the VRS output to produce reliable comparator triggering on each pass.  Record the VRS setting and the interrupter output voltage with and without the beam blocked.

VRS output _____ V

Interrupter output, unblocked _____ V

Interrupter output, blocked _____ V

Measure the voltages at the TL082 + input when the comparator output is HI and LO.

+ input, 311 HI _____ V

+ input, 311 LO _____ V

Question 11.  Calculate the + input voltages for HI and LO output expected for the circuit and compare with the observed values.  How much hysteresis does this circuit provide?

Connect the CTFM A input to the optointerrupter output.  Set the CTFM to the unit count mode, and reset the counter.  Use an opaque card to block and unblock the beam.  Record the average count for five insertion-removal cycles.

Average counts _____

Now move the CTFM A input to the comparator output and again insert and remove the card. Record the number of counts per insertion/removal cycle. Is the count advanced on the insertion or removal?

Counts/cycle _____

Advanced on _____ edge

Count the number of pins on an IC by pulling it through the interrupter. If the count produced is not accurate or reproducible, readjust the VRS threshold source. Obtain an IBM card with a number of holes punched in the 9-row, and count these holes. Attempt to determine the maximum hole count rate. Report your results.

Leave the Schmitt trigger circuit connected for the next experiment.

## 9-3 CHARGE-TO-COUNT AND VOLTAGE-TO-FREQUENCY CONVERTERS

Objectives: To investigate the design and characteristics of voltage-to-frequency converters and their use in the charge-to-count converter and the integrating DVM by constructing a VFC from previously studied components and observing its response and characteristics.

The voltage-to-frequency converter (VFC) will be constructed using the integrator, Schmitt trigger, 555-type monostable and analog switch circuits as shown in figure 9-6. Use a carefully balanced op amp for the integrator.

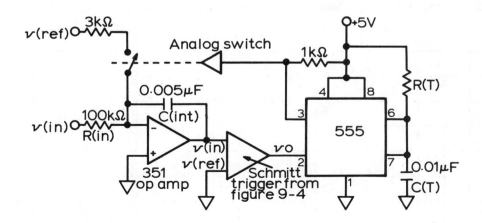

Fig. 9-6

156

Note that the 555 circuit differs in only one component from the astable multivibrator of experiment 9-1, and that the Schmitt is identical to that constructed in experiment 9-2. Connect scope Ch 1 to the 555 timer output, and Ch 2 to the integrator output. Trigger on the positive edge of the 555 output. Set v(in) to -10 V and v(ref) to +0.8 V. Connect the CTFM A input to the 555 timer output. Set the CTFM in the frequency mode with a 1.0 s gate time. The scope should display a triangular waveform from the integrator, and a pulse waveform from the 555 output. If not, recheck the wiring, particularly the sign and magnitude of v(ref) and v(in).

Connect the DMM to monitor v(in). Calibrate the VFC by adjusting v(ref) until the CTFM reading in kilohertz equals the DMM reading in volts. Record the value of v(ref) at calibration.

v(ref) at calibration _____ V

Record the duration of the HI portion of the 555 timer output and the VFC output frequency measured on the CTFM at v(in) = -10.0, -5.00 V and -1.00 V

| v(in) | Output Duration | CTFM Frequency |
|-------|-----------------|----------------|
| -10.0 V | _____ | _____ |
| - 5.0 V | _____ | _____ |
| - 1.0 V | _____ | _____ |

If the output duration is not constant, the integrating capacitor is probably too large, Replace with a smaller value, recalibrate the VFC, and repeat the above measurements.

Set v(in) to -10 V. Sketch the waveforms at the integrator output and 555 output in sufficient detail that both the positive and negative slopes of the integrator output can be calculated. Show at least one full cycle of each display. Leave room below each sketch for a third waveform. Move the Ch 2 scope input from the integrator output to the comparator output. Add this waveform to the other two, carefully observing the time relationship. Indicate the portion of the waveform during which the switch is on.

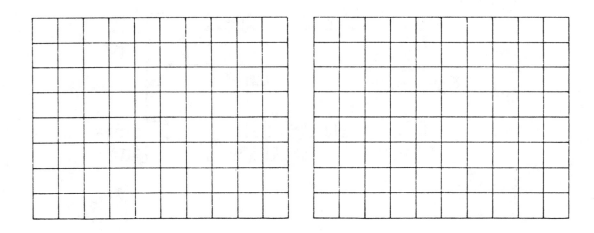

Sketch the integrator output and 555 timer output as above for v(in) = -5 V and v(in) = - 200 mV.

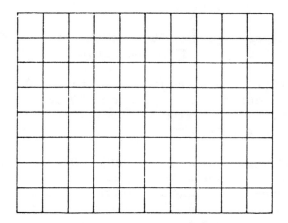 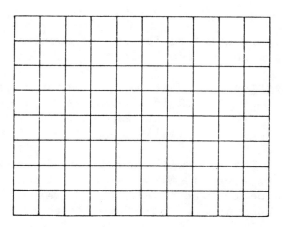

Question 12. From the slope of the integrator output when the analog switch is off, calculate the charge input to the integrator from v(in) during one entire cycle of the VFC output. Do this for all three input voltages. Also calculate the total charge input during the time the switch is on for each value of v(in). Compare the charge input from v(ref) to the charge input from v(in) during a complete cycle of the VFC, and account for any discrepancies. What is the current input from v(ref) during the switch on time? Compare with the values calculated from component values and comment on any differences.

Question 13. What range of values could be used for C(int) without changing other components, taking into consideration the hysteresis of the comparator and the voltage limitation of the integrator? Within this range, would changing the value of C(int) affect the calibration?

Question 14. Specify values for Rt, Ct and C(int) for operation of this circuit at 100 kHz full scale.

Measure the voltage v(in) and the VFC frequency out for 6 different voltages spanning at least three decades. Use a 1 s frequency measurement time.

| v(in), V | VFC out, Hz | Ratio, Hz/V |
|----------|-------------|-------------|
| _____ | _____ | _____ |
| _____ | _____ | _____ |
| _____ | _____ | _____ |
| _____ | _____ | _____ |
| _____ | _____ | _____ |
| _____ | _____ | _____ |

Question 15. In cases where there are discrepancies between the VFC and the DMM, which do you believe and why?

Change R(in) from 100 kΩ to 10 kΩ. What voltage at v(in) now produces the 10 kHz output?

10 kHz output from _____ V input

Restore R(in) to 100 kΩ and continue to increase v(in). Observe the effect of too high an input voltage on the output, both via the CTFM and scope.

Question 16. What is the input current that produces 10 kHz with a 10 kΩ R(in). Compare with the input current that produces 10 kHz with a 100 kΩ resistor. Comment on the statement that the VFC is really a current-to-frequency converter.

The VFC will now be used as a charge-to-count converter. Disconnect v(in) from the VFC. Wire a jumper arrangement such that a 1.00 μF capacitor can be charged to −10 V with respect to ground and then discharged to the VFC through R(in). Set the CTFM to the unit counting mode. Charge the capacitor to −10 V. Reset the CTFM. Discharge the capacitor through the VFC. Record the average count and precision from five trials. Repeat this experiment for three other capacitance values spread over three or four decades.
Repeat the experiment with an unknown capacitor obtained from the instructor and record the voltage to which the capitor charges as measured by the DMM.

| Capacitor | Count | Precision |
|-----------|-------|-----------|
| _____ | _____ | _____ |
| _____ | _____ | _____ |
| _____ | _____ | _____ |
| _____ | _____ | _____ |
| _____ | _____ | _____ |

Charge voltage _____ V

Leave the VFC connected for the next experiment.

Question 17. Calculate from the Count values in the table above, the measured capacitance of the devices. Compare (where possible) with nominal values and discuss any deviations.

The next experiments illustrate the signal averaging characteristics of the integrating DVM made from the VFC and the CTFM. The noise generator introduced in Unit 8 will be used to show random noise reduction through averaging. The noise generator and summing amplifier are located on the REF job board.

Set the noise amplitude to zero volts with the amplitude adjust potentiometer (see Figure 8-13). Connect the VRS $\pm$ 10 V output to v(in) of the noise mixer (NM) and adjust the VRS for -5 V at the NM output. Connect the NM output to the VFC input and record the frequency reading for this noise-free signal. Use a 1 s time base on the CTFM.

CTFM reading _____.

Display on the oscilloscope the VFC 555 timer output and the VFC input signal with a 50 µs/div time base. Trigger on the + slope of the 555 output. Increase the noise amplitude to approximately 1 V p-p. Sketch the 1 V p-p waveforms observed.

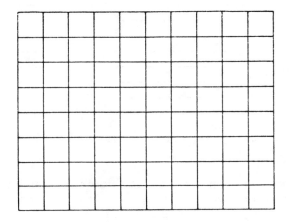

Observe the degree of variation in 555 pulse timing with the introduction of noise. Record the VFC output frequency and estimate its uncertainty (f(max)-f(min)) for all CTFM time bases available. Repeat for 4 V of noise, p-p.

| Gate Interval | Frequency | counter uncertainty 1 V noise | uncertainty 4 V noise |
|---|---|---|---|
| 0.01 s | _____ | _____ | _____ |
| 0.1 s | _____ | _____ | _____ |
| 1.01 s | _____ | _____ | _____ |
| 10.0 s | _____ | _____ | _____ |

Question 18. Do the values for the uncertainty follow the predicted effects of increasing integration time? Compare the observed effect of the noise amplitude with that expected.

The ability of an integrating DVM to reject fixed-frequency noise depends upon the relation of the noise frequency to the integration interval. Connect the FG sine wave output to the VFC input. Set the output to 4 V p-p and offset by -5 V. The FG frequency will have to be set very accurately in this experiment, so make it convenient to alternately measure the FG TTL output period and the VFC output frequency.

Set the FG frequency as accurately as you can to the values below and record the maximum and minimum VFC output frequency observed over a 10-15 s interval. For VFC output frequency measurements, use a 1.0 s gate interval.

| frequency | minimum reading | maximum reading |
|-----------|-----------------|-----------------|
| 1.0 Hz | _____ | _____ |
| 1.5 Hz | _____ | _____ |
| 2.0 Hz | _____ | _____ |
| 2.5 Hz | _____ | _____ |
| 3.0 Hz | _____ | _____ |
| 3.5 Hz | _____ | _____ |
| 5.0 Hz | _____ | _____ |
| 5.5 Hz | _____ | _____ |

Graph 1. Plot the uncertainty of the measurement (maximum) vs. frequency. Explain your results.

Question 19. Comment on the importance of exact frequency adjustment for the effective rejection of noise. Comment on the effectiveness of 0.01 s, 0.1 s, and 1 s gate intervals on the rejection of 60 Hz power line noise. Would a European with 50 Hz power be better off?

9-4   DUAL SLOPE CONVERTER

Objectives:   To study the operation and characteristics of a dual-slope analog-to-digital converter by experimenting with an operating LSI based digital panel meter and observing its measurement sequence and response.

In this experiment, the operation and characteristics of a dual-slope DVM IC are studied. The IC selected (the Intersil 7106) for this experiment is typical of those used in digital panel meters and digital multimeters. The Intersil 7106 is mounted on a circuit board with passive components and display to make a complete $\pm$ 2V DVM.

Turn on the DVM. Carefully follow the instructions marked on the unit. Connect the VRS $\pm$ 1 V output to both the DVM and the DMM. Compare readings of the two instruments for several voltages over the range of the VRS. If the two meters do not agree, do not attempt a calibration without first consulting your instructor. Record your accuracy checks below.

| Positive v(in) | | Negative v(in) | |
|---|---|---|---|
| DVM | DMM | DVM | DMM |
| ___ | ___ | ___ | ___ |
| ___ | ___ | ___ | ___ |
| ___ | ___ | ___ | ___ |

The analog section of the 7106 DVM IC is shown in figure 9-7.

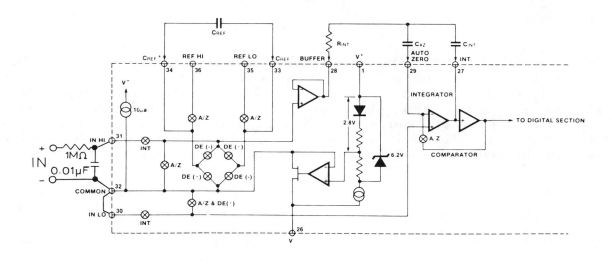

Fig. 9-7

The input signal is connected to the buffer amplifier through the INT switch during the signal integration interval. Capacitor C(ref) has been charged to the reference voltage. At the end of the input signal integration interval, the switching bridge labelled DE connects C(ref) to the buffer amplifier. The polarity of this connection is determined by the output of a signal polarity sensor (not shown) so that the reference integration is always opposite to that of the signal. The switches marked A/Z are closed after the reference integration and before the next signal integration. These switches are part of the auto-zero system that charges C(AZ) to the combined offset voltages of the buffer, integrator and comparator amplifiers. The integrator summing point is then offset to this value during the next integrating cycle. This automatically compensates for offsets and drifts on each measurement cycle.

Connect scope Ch 1 to the buffer output test point (TP6). The test point locations are shown in figure 9-8.

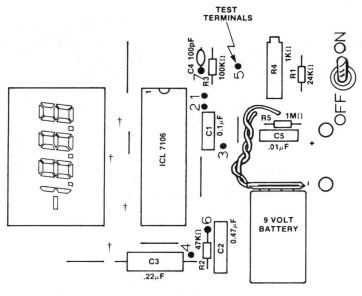

Fig. 9-8

Connect scope Ch 2 to the integrator output test point (TP4). Apply a signal of +0.5 V to the DVM input and set the scope time base to 50 ms/div. The Ch 1 trace should display a positive pulse of 0.5 V amplitude followed immediately by a pulse of −1.0 V. Adjust the scope to trigger on the positive-going edge at the end of the negative pulse. Set Ch 1 input to 0.5 V/div and Ch 2 to 200 mV/div. Sketch the waveforms observed. Change the input voltage to +1.0 V and sketch the waveforms again. Record the exact DVM reading.

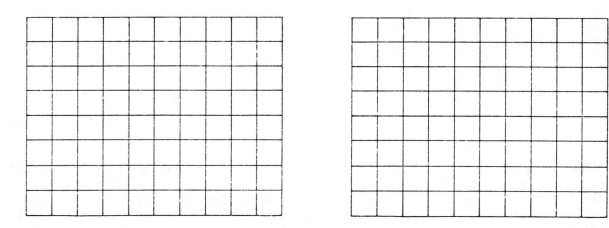

Question 20. Identify the period during which the signal is being integrated and the period of integration of the reference signal. Does the signal integration interval change with signal level? Does the reference integration time change? How is the slope of the integrator output during signal and reference integration times affected by the input signal voltage? Compare the areas under the positive and negative pulses for each input voltage.

Change the input voltage to −0.5 V to observe the change in the waveforms for negative signals. Trigger the scope on the falling edge of the reference voltage pulse at the buffer output (TP6). Sketch the waveforms and record the DVM reading. Vary the input voltage over a range of negative values and observe the effect on the waveforms.

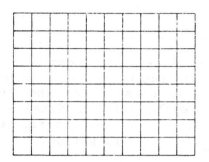 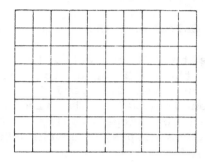

<u>Question 21.</u>    Compare the areas under the reference and signal integration pulses in your sketches above.

Connect the CTFM input to the oscillator test point (TP7) on the DVM board. Measure the DVM oscillator frequency. The DVM oscillator goes through a ÷4 scaler before being connected to the DVM display counter. What is the display count rate?

Oscillator frequency _____ kHz

Display count rate    _____ kHz

<u>Question 22.</u>    From the waveforms obtained earlier and the measured count rate, calculate the number of counts that occur during the signal integration interval. Also compare the calculated number of counts during the reference integration interval with the DVM reading for the voltages measured during the waveform observations.

The response of the dual-slope converter to fixed pattern noise is tested in the experiment below. Set the FG signal output for 1.5 V p-p, 20 Hz sine wave centered on −1.0 V dc. Connect this signal to the DVM input. With the same scope connections used before, sketch the buffer and integrator output waveforms. Note the 20 Hz oscillations in the input signal appear during the signal integration period at the buffer output. The 20 Hz component of the buffer output is reduced in amplitude relative to the FG output. Note also the non-linearity of the integrator output during the integration of the signal.

Question 23. Consider the RC filter in the DVM input circuit and explain the observed attenuation of the 20 Hz signal.

Calculate the signal integration interval of the DVM exactly from the CFTM measurement of the DVM clock and the exact number of clock cycles in the integration interval (4000).

Calculated integration time _____ ms

Leave the FG connected to the DVM as above. Also connect the FG TTL output to the CTFM input to measure the FG output period. Adjust the FG frequency so that its period is as close as possible to the time of the integration interval. The single "noise" cycle that is integrated can be observed in the scope display of the buffer output. Record the minimum and maximum values displayed by the DVM over a 10-15 second period.

DVM, minimum _____

DVM, maximum _____

Calculate the frequencies of maximum rejection between 10 and 100 Hz.

Rejection frequencies: _____, _____, _____,

_____, _____, _____, _____, _____,

_____, _____.

Pick the frequency closest to 60 Hz and observe the noise rejection at that frequency. Also determine the noise rejection at frequencies 5% above and below the "nearest 60 Hz" rejection node and also at exactly 60 Hz.

| | nearest 60 Hz | | | |
|---|---|---|---|---|
| | node | +5% | -5% | exactly 60 Hz |
| DVM, maximum | _____ | _____ | _____ | _____ |
| DVM, minimum | _____ | _____ | _____ | _____ |

Question 24. Discuss the line frequency rejection characteristics of your DVM. How could the line rejection characteristics of your DVM be improved?

Question 25. The VFC integrating DVM has a problem with noisy signals when the noise component causes a "zero" crossing of the signal voltage, i.e. when the signal momentarily changes polarity. Explain why this is the case. What effect does such polarity-reversing noise have on the reading of the dual slope DVM? Explain.

9-5  PHASE-LOCKED LOOP

Objectives:  To investigate a phase-locked loop IC by determining the free-running frequency of its internal voltage-controlled oscillator, by determining the frequency range of lock and by constructing a frequency multiplier.

Locate the 565 phase-locked loop (PLL) IC on the advanced analog circuits job board.  The pinout and block diagram of the PLL are given in figure 9-9.

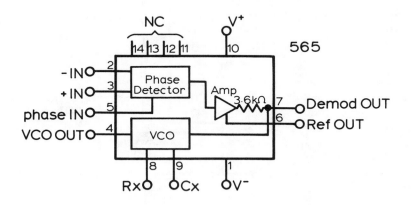

Fig. 9-9

The 565 PLL operates over a power supply voltage range of $\pm$ 5 V to $\pm$ 12 V.  In this experiment it will be operated from $\pm$ 5 V supplies.  To generate the -5 V supply voltage, a -5 V regulator will be used to drop the -15 V power supply voltage to -5 V.  Locate the 7905 regulator on the advanced analog circuits job board near the 565 PLL.  All connections to the regulator have been prewired.  Verify that the output voltage of the regulator is $\simeq$ -5 V by measuring the output voltage as shown in figure 9-10 with the DMM.

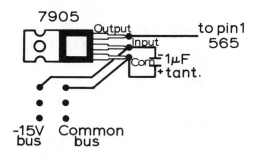

Fig. 9-10

Regulator output voltage _____

Now the PLL circuit of figure 9-11 will be used to investigate the PLL characteristics. Wire the circuit, but do not connect the FG input yet. Timing components Rx and Cx determine the free running frequency fo of the VCO according to fo ~ 1/(3.7 RxCx).

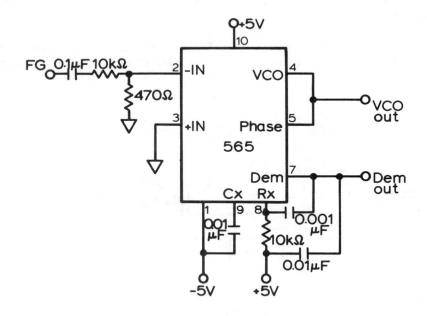

Fig. 9-11

The 0.01 μF capacitor on the demodulator output in conjunction with the internal 3.6 kΩ resistor form a low pass filter to filter the amplified output of the phase detector (multiplier). With the FG still disconnected observe the free running VCO output on the oscilloscope. Measure the p-p output voltage.

VCO output voltage (p-p) _____

Connect the CTFM to the VCO output simultaneously with the scope, and measure the free running frequency fo. Compare it to that expected.

fo (measured) _____

fo (expected) _____

Now connect the FG sine wave output to the PLL input through the divider network shown in figure 9-11. The capacitor removes any dc component of the sine wave, while the divider allows a small input voltage to be applied to the PLL while a larger voltage is available for scope observation. Connect scope Ch 1 to the FG output and Ch 2 to the VCO output. (Trigger the scope on the FG output). Adjust the FG sine wave amplitude (before the divider) to the ~ 6 V peak-to-peak. Set the FG frequency to be near the free running frequency of the VCO. Vary the FG output frequency. The loop is locked when the VCO output frequency follows the FG frequency.

Locking should be observed over a frequency range of approximately a factor of 2. To determine the lock range accurately, connect the FG TTL output to the A input of the CTFM and the VCO output of the PLL to the B input. (The FG sine wave output should still be connected to the PLL). When the loop is locked the frequency ratio of A/B should be exactly 1.00. By switching the CTFM from frequency ratio mode to frequency mode, determine the range of input frequencies over which the loop is locked.

Lock range _____ Hz to _____ Hz

Disconnect the CTFM and connect the scope to observe the PLL input (Ch 1) and the VCO output (Ch 2). Adjust the FG frequency to be in the center of the lock range. Determine the phase shift between the FG input and the VCO output near the center of the lock range.

Phase shift _____

Question 26. Discuss why a phase shift near that observed is necessary for the loop to be locked.

Now observe on the scope the following pairs of outputs: (1) FG and VCO; (2) FG and Cx (pin 9); (3) FG and demodulator (pin 7). Plot the VCO, the timing capacitor and demodulator outputs relative to the FG waveform on the graph paper below.

Now connect the DMM to the demodulator output while observing the FG output and the VCO output on the scope. Measure the demodulator output voltage for 5 or 6 FG input frequencies while the loop is locked. Measure the FG frequency on the CTFM.

FG freq.          Demodulator output
_____           _____
_____           _____
_____           _____
_____           _____
_____           _____
_____           _____

Observe what happens to the demodulator output voltage when the loop goes out of lock at both ends of the lock range. While the loop is unlocked, shut off the FG and observe the demodulator output on the DMM.

Dem output (FG on, loop unlocked) _____

Dem output (FG off)                _____

Question 27.  Discuss how the PLL could be used as an FM demodulator or a frequency-to-voltage converter.

Next the PLL will be used to make a frequency multiplier. A divide by 10 counter will be inserted into the loop between the VCO output and the phase detector input. The VCO output should then be exactly 10 X the input frequency supplied from the FG. Break the loop between the VCO output (pin 4) and the phase detector input (pin 5). Make no other changes. Connect the divider circuit shown in figure 9-12 between the VCO output and the phase detector input. The counter and the 7404 inverters are located on the counter job board. The inverters are used to sharpen the VCO output transitions (two are used to keep the phase correct). The 7490 counter is connected as a symmetrical divide by 10 counter instead of its usual BCD counter mode (divide by 2 then divide by 5).

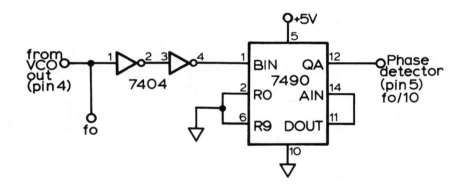

Fig. 9-12

Set the FG output to ~ 1/10 the free running frequency of the VCO. Observe on the scope the FG sine wave output and the VCO output. When lock is obtained, connect the VCO output to the A input of the CTFM and the FG output to the B input. Obtain the frequency ratio in the lock range.

Ratio of VCO output freq. to FG output freq. _____

As before, determine the lock range.

Range of FG frequencies over which lock is maintained _____

Question 28.  Design a circuit that would produce switch selectable output frequencies of 2 kHz, 10 kHz and 40 kHz from a 1 kHz oscillator. Use only a PLL, a 7473 dual JK flip flop and a 7490. Will the PLL free running frequency have to be varied?

# UNIT 10. LOGIC GATES, FLIP-FLOPS, AND COUNTERS

The NAND, NOR, and Exclusive-OR logic gates in the famous 7400 series TTL family of integrated circuits are studied first in these experiments. Various combinations of these basic inverting logic gates are shown to provide the important comparator, adder, multiplexer and decoder operations. Basic gates are also combined to form latches and flip-flops, and the IC versions of these devices are studied. Flip-flops are combined to form counters and an IC up/down counter is connected and operated in conjunction with a versatile counter input gate. The shift register, studied in the final experiment, is shown to function as a serial-to-parallel converter, a parallel-to-serial converter, and a shift left/shift right register. These experiments continue the study of basic logic devices that was begun in Unit 4.

|    Equipment    |    Parts    |
|-----------------|-------------|

1. Basic station
2. CTFM
3. BSI job board
4. Reference job board
5. Basic gates job board
6. MSI gates job board
7. Flip-flop job board
8. Counter job board

## 10-1 BASIC LOGIC GATES

Objectives: To introduce the inverting gates, NAND, NOR, and AND-OR-INVERT (AOI) and to demonstrate how equivalent logic functions can be achieved with either NAND or NOR gates by observing the tables of states for the inverting gates and the equivalent tables of states for equivalent functions.

Install the basic gates job board in the frame. The integrated circuits are arranged on the job board in the order shown in figure 10-1.

170

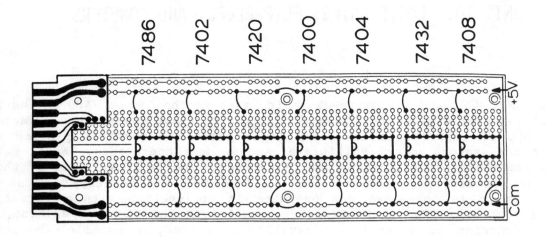

Fig. 10-1

There are two NAND gates on the basic gates job board. The 7400 chip contains four, 2-input NAND gates, while the 7420 has two, 4-input gates as shown by the pinouts in figure 10-2.

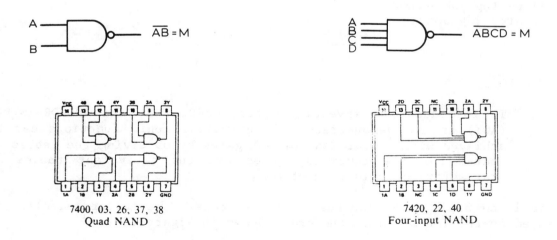

7400, 03, 26, 37, 38
Quad NAND

7420, 22, 40
Four-input NAND

Fig. 10-2

Use logic sources for inputs and an LED indicator for the output, and verify the table of states for one gate on each chip and report the results in the table provided. Since this is a table of states for a device, the data should be recorded as LO and HI, not as 1 and 0. The latter is an arbitrary assignment not inherent in the device. (See note 10-1, page 254, in the text).

| Two input gate | | |
| --- | --- | --- |

| Inputs | | Output |
| --- | --- | --- |
| A | B | M = $\overline{AB}$ |
| L | L | _____ |
| L | H | _____ |
| H | L | _____ |
| H | H | _____ |

| Four input gate | | | | |
| --- | --- | --- | --- | --- |

| Inputs | | | | Output |
| --- | --- | --- | --- | --- |
| A | B | C | D | M = $\overline{ABC}$ |
| ___ | ___ | ___ | ___ | _____ |
| ___ | ___ | ___ | ___ | _____ |
| ___ | ___ | ___ | ___ | _____ |
| ___ | ___ | ___ | ___ | _____ |
| ___ | ___ | ___ | ___ | _____ |
| ___ | ___ | ___ | ___ | _____ |
| ___ | ___ | ___ | ___ | _____ |
| ___ | ___ | ___ | ___ | _____ |
| ___ | ___ | ___ | ___ | _____ |
| ___ | ___ | ___ | ___ | _____ |
| ___ | ___ | ___ | ___ | _____ |
| ___ | ___ | ___ | ___ | _____ |
| ___ | ___ | ___ | ___ | _____ |
| ___ | ___ | ___ | ___ | _____ |
| ___ | ___ | ___ | ___ | _____ |
| ___ | ___ | ___ | ___ | _____ |

Question 1.   For the 2-input NAND gate, rewrite the table of states observed using 1 for LO and 0 for HI.  This is now the truth table for the logic function performed by the gate on LO-true signals. Identify the function.

Similarly investigate the table of states for the 7402 quad, 2-input NOR gate shown in figure 10-3

7402, 28, 33
Quad NOR

Fig. 10-3

| Inputs | | Output |
| --- | --- | --- |
| A | B | M = $\overline{A+B}$ |
| ___ | ___ | _____ |
| ___ | ___ | _____ |
| ___ | ___ | _____ |
| ___ | ___ | _____ |

Question 2.   For the 2-input NOR gate, write the truth table (1's and 0's) for the response observed for LO-true signals.  Identify the function.

<u>Question 3</u>. If a NAND or NOR gate is available and the inverter function is needed, how would you connect the gate to achieve this function?

<u>Question 4</u>. A seat belt warning system is to sound a warning buzzer unless all passengers have their seat belts fastened. The warning buzzer is to sound unless the driver's belt is fastened <u>and</u> each of the other seat belts are fastened <u>or</u> there is no weight on the seat. Assume there are HI true logic signals representing the following: SS = start switch ON, SB1 = driver seat belt fastened, SB2, SB3, SB4 = other belts fastened, W2, W3, W4 = weight sensed on seats 2, 3 and 4. Write a logic equation for buzzer ON in terms of the above signals. Draw a diagram to show how this equation can be implemented with the gates studied in this experiment. Identify the gates used by their 7400-series number.

All the basic gate functions can be performed by one type of inverting gate. Often it is convenient or economical to use one type of gate for all the required logic functions. Show experimentally, using the basic gates job board, that the three AND gate circuits in figure 10-4 are equivalent and the two OR gate circuits in figure 10-5 are equivalent. Logical equivalence is demonstrated by identical tables of states. Record your data in the tables.

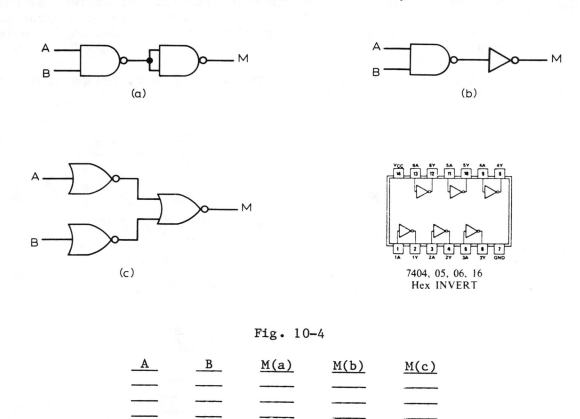

Fig. 10-4

| A | B | M(a) | M(b) | M(c) |
|---|---|------|------|------|
| — | — | ——— | ——— | ——— |
| — | — | ——— | ——— | ——— |
| — | — | ——— | ——— | ——— |
| — | — | ——— | ——— | ——— |

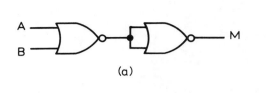

(a)

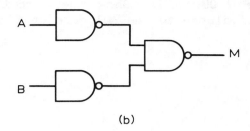

(b)

Fig. 10-5

| A | B | M(a) | M(b) |
|---|---|------|------|
| — | — | — | — |
| — | — | — | — |
| — | — | — | — |
| — | — | — | — |

Figure 10-6 shows an AND-OR-INVERT (AOI) gate.  Implement this with two AND gates and one NOR gate.  Record the table of states.

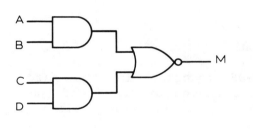

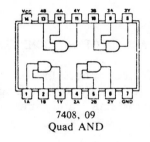

7408, 09
Quad AND

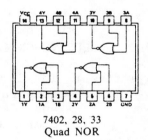

7402, 28, 33
Quad NOR

Fig. 10-6

| A | B | C | D | M |
|---|---|---|---|---|
| — | — | — | — | — |
| — | — | — | — | — |
| — | — | — | — | — |
| — | — | — | — | — |
| — | — | — | — | — |
| — | — | — | — | — |
| — | — | — | — | — |
| — | — | — | — | — |
| — | — | — | — | — |
| — | — | — | — | — |
| — | — | — | — | — |
| — | — | — | — | — |
| — | — | — | — | — |
| — | — | — | — | — |
| — | — | — | — | — |
| — | — | — | — | — |

The AOI gate can be implemented with only NAND gates. Implement it with one 7400 chip. Show the circuit and table of states below to prove its equivalence to the AOI gate of figure 10-6.

10-2   EXCLUSIVE-OR GATE AND DIGITAL COMPARATORS

Objectives:   To investigate the Exclusive-OR function in its gate and comparator forms by observing the tables of states of the 7486 and 7485 devices and relating these to the desired logic operations.

The pinout of the 7486 quad, 2-input Exclusive-OR gate is shown in figure 10-7.  Determine the table of states for one of the gates.  Record your data as LO and HI (or L and H).

7486
Exclusive-OR

Fig. 10-7

| Inputs | | Output |
| --- | --- | --- |
| A | B | M = A ⊕ B |
| ____ | ____ | _____ |
| ____ | ____ | _____ |
| ____ | ____ | _____ |
| ____ | ____ | _____ |

Question 5. Draw an all NAND gate implementation of the Exclusive-OR function.

The Equality or coincidence function can be implemented by inverting the output of the Exclusive-OR gate. Implement this with one 7486 gate and one 7404 inverter. Fill in the table of states below.

| Inputs | | Output |
| --- | --- | --- |
| A | B | M = A ⊕ B |
| ____ | ____ | _____ |
| ____ | ____ | _____ |
| ____ | ____ | _____ |
| ____ | ____ | _____ |

Connect the four-bit equality detector or digital comparator shown in figure 10-8. Implement the 4 equality gates with Exclusive-OR gates and inverters. Use logic level sources SA-SD for word A and SE-SH for word B. Source SA should be the most significant bit of word A in order for the switch position to mimic the binary word written left to right. Show that the equality condition is indicated at the output only when all four bits of the two digital words are equal. Try the combinations suggested in the table and several others.

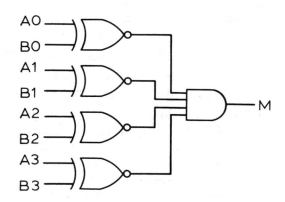

Fig. 10-8

| Word A | | | | Word B | | | | Output |
|---|---|---|---|---|---|---|---|---|
| A3 | A2 | A1 | A0 | B3 | B2 | B1 | B0 | M |
| L | H | H | L | L | H | H | L | ___ |
| L | H | L | L | L | H | H | L | ___ |
| L | H | L | L | L | H | H | L | ___ |
| ___ | ___ | ___ | ___ | ___ | ___ | ___ | ___ | ___ |
| ___ | ___ | ___ | ___ | ___ | ___ | ___ | ___ | ___ |
| ___ | ___ | ___ | ___ | ___ | ___ | ___ | ___ | ___ |
| ___ | ___ | ___ | ___ | ___ | ___ | ___ | ___ | ___ |

The 7485 magnitude comparator is a medium scale integrated circuit (MSI chip). Its pinout is shown in figure 10-9. It performs magnitude comparison of binary and BCD coded digital words.

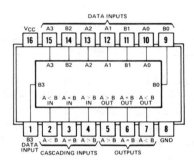

Fig. 10-9

The 7485 comparator is located on the MSI gates job board. Connect the A inputs to 4 logic level sources and the B inputs to 4 more logic level sources. Connect the A > B, A < B and A = B outputs to indicator lights. Use several A and B input combinations and fill in the table of states below.

| A3 | A2 | A1 | A0 | B3 | B2 | B1 | B0 | A > B | A < B | A = B |
|---|---|---|---|---|---|---|---|---|---|---|
| H | L | H | H | H | L | H | L | ___ | ___ | ___ |
| H | L | H | L | H | L | H | H | ___ | ___ | ___ |
| H | H | L | L | H | H | L | L | ___ | ___ | ___ |
| ___ | ___ | ___ | ___ | ___ | ___ | ___ | ___ | ___ | ___ | ___ |
| ___ | ___ | ___ | ___ | ___ | ___ | ___ | ___ | ___ | ___ | ___ |
| ___ | ___ | ___ | ___ | ___ | ___ | ___ | ___ | ___ | ___ | ___ |
| ___ | ___ | ___ | ___ | ___ | ___ | ___ | ___ | ___ | ___ | ___ |

Question 6. Describe the logic in the comparator chip that is needed to obtain the A = B output, the A > B output and the A < B output. Use diagrams and Boolean expressions in your description.

## 10-3 LOGIC OPERATIONS

Objectives: To explore the operation and implementation of the medium-complexity logic operations of adding, decoding, and multiplexing by wiring these operations with combinations of basic gates and with special-purpose MSI logic IC's.

For this experiment, install both the basic gates job board and the MSI gates job board. Wire the half-adder shown in figure 10-10 using an AND gate and an Exclusive-OR gate on the basic gates job board. Complete the table of states below. Show that the circuit produces the desired logic function for HI-true signals.

| Inputs | | Outputs | |
| --- | --- | --- | --- |
| | | Sum | Carry |
| A | B | $A \oplus B$ | $A \cdot B$ |
| ____ | ____ | _____ | _____ |
| ____ | ____ | _____ | _____ |
| ____ | ____ | _____ | _____ |
| ____ | ____ | _____ | _____ |

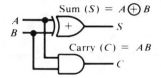

Fig. 10-10

Locate the 7483 full adder chip on the MSI gates job board and refer to the pinout in figure 10-11. This adder produces a four-bit sum ($\Sigma$) and a carry out (CO) from 2, four-bit words (A and B) and a carry in (CI).

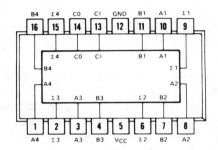

Fig. 10-11

The table below describes the binary addition of A = 10 and B = 9 to give a sum = 19. In the pin out the least significant bit (LSB) of a word A is designated A1, the next most significant bit A2, etc.

|  | MSB |  |  | LSB |  |
|---|---|---|---|---|---|
|  | 4 | 3 | 2 | 1 | CI |
| Word A | 1 | 0 | 1 | 0 | 0 |
| Word B | 1 | 0 | 0 | 1 | 0 |
| Sum | 0 | 0 | 1 | 1 |  |
| Carry out, CO 1 |  |  |  |  |  |

Use switches SA-SD to generate logic level signals to represent word A and switches SE-SH for word B. Wire the switch outputs to the adder inputs according to the pin diagram. Connect the carry in to common to provide CI = 0. Connect the sum outputs ( 1 - 4) and the carry out to LED indicators. Perform the binary addition of at least five pairs of 4-bit binary numbers and record the results in the table below. The designated function is performed for HI-true signals only.

| Word A | | | | Word B | | | | Sum | | | | Carry out | Decimal Equivalent |
|---|---|---|---|---|---|---|---|---|---|---|---|---|---|
| A4 | A3 | A2 | A1 | B4 | B3 | B2 | B1 | 4 | 3 | 2 | 1 | CO | |
| __ | __ | __ | __ | __ | __ | __ | __ | __ | __ | __ | __ | __ | __ |
| __ | __ | __ | __ | __ | __ | __ | __ | __ | __ | __ | __ | __ | __ |
| __ | __ | __ | __ | __ | __ | __ | __ | __ | __ | __ | __ | __ | __ |
| __ | __ | __ | __ | __ | __ | __ | __ | __ | __ | __ | __ | __ | __ |
| __ | __ | __ | __ | __ | __ | __ | __ | __ | __ | __ | __ | __ | __ |

Enter another combination of two words A and B and obtain the results for CI = 0 (Common) and CI = 1 (+5 V). Record your results below.

Question 7. Sketch a block diagram of a circuit that uses 7483's to add two 8-bit words.

Question 8. Two nuclear counters are used, one to count β particles and the other γ particles. Parallel digital outputs are available from each counter. A warning signal is desired when the sum of the β and γ particles exceeds a preset binary number. Sketch a block diagram of a circuit to produce the warning signal.

A binary decoder provides a separate logic level output for all combinations of input logic levels. The binary decoder of figure 10-12 decodes the four possible states of the two input signals E and F into separate output lines Q, R, S and T and is therefore called a 2-line to 4-line binary decoder. Wire the circuit of figure 10-12 on the basic gates job board. Verify the decoder function by filling out the following table of states.

| Inputs | | Outputs | | | |
|---|---|---|---|---|---|
| E | F | Q | R | S | T |
| L | F | ___ | ___ | ___ | ___ |
| L | H | ___ | ___ | ___ | ___ |
| H | L | ___ | ___ | ___ | ___ |
| H | H | ___ | ___ | ___ | ___ |

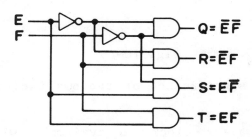

Fig. 10-12

Locate the 7442 BCD-to-decimal decoder on the MSI gates job board. Refer to the pinout of figure 10-13 and connect four logic sources to the A-D inputs of the chip. Connect each decoder output to an LED indicator. Construct a table of states for all input combinations.

| Inputs | | | | Outputs | | | | | | | | | |
|---|---|---|---|---|---|---|---|---|---|---|---|---|---|
| A | B | C | D | 0 | 1 | 2 | 3 | 4 | 5 | 6 | 7 | 8 | 9 |
| ___ | ___ | ___ | ___ | ___ | ___ | ___ | ___ | ___ | ___ | ___ | ___ | ___ | ___ |
| ___ | ___ | ___ | ___ | ___ | ___ | ___ | ___ | ___ | ___ | ___ | ___ | ___ | ___ |
| ___ | ___ | ___ | ___ | ___ | ___ | ___ | ___ | ___ | ___ | ___ | ___ | ___ | ___ |
| ___ | ___ | ___ | ___ | ___ | ___ | ___ | ___ | ___ | ___ | ___ | ___ | ___ | ___ |
| ___ | ___ | ___ | ___ | ___ | ___ | ___ | ___ | ___ | ___ | ___ | ___ | ___ | ___ |
| ___ | ___ | ___ | ___ | ___ | ___ | ___ | ___ | ___ | ___ | ___ | ___ | ___ | ___ |
| ___ | ___ | ___ | ___ | ___ | ___ | ___ | ___ | ___ | ___ | ___ | ___ | ___ | ___ |
| ___ | ___ | ___ | ___ | ___ | ___ | ___ | ___ | ___ | ___ | ___ | ___ | ___ | ___ |
| ___ | ___ | ___ | ___ | ___ | ___ | ___ | ___ | ___ | ___ | ___ | ___ | ___ | ___ |
| ___ | ___ | ___ | ___ | ___ | ___ | ___ | ___ | ___ | ___ | ___ | ___ | ___ | ___ |
| ___ | ___ | ___ | ___ | ___ | ___ | ___ | ___ | ___ | ___ | ___ | ___ | ___ | ___ |
| ___ | ___ | ___ | ___ | ___ | ___ | ___ | ___ | ___ | ___ | ___ | ___ | ___ | ___ |
| ___ | ___ | ___ | ___ | ___ | ___ | ___ | ___ | ___ | ___ | ___ | ___ | ___ | ___ |
| ___ | ___ | ___ | ___ | ___ | ___ | ___ | ___ | ___ | ___ | ___ | ___ | ___ | ___ |
| ___ | ___ | ___ | ___ | ___ | ___ | ___ | ___ | ___ | ___ | ___ | ___ | ___ | ___ |
| ___ | ___ | ___ | ___ | ___ | ___ | ___ | ___ | ___ | ___ | ___ | ___ | ___ | ___ |

180

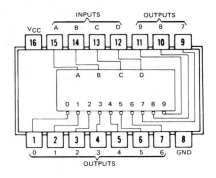

Fig. 10-13

Question 9. What is the significance of the small circle on each output of the pinout?

Question 10. A digital alarm clock is available with four lines which indicate unit hours in BCD code. If these outputs are connected to the A-D inputs of the 7442 decoder and decoder outputs 1 and 5 are connected to the inputs of a NAND gate, predict the hours of the day that the NAND gate output will be HI.

Digital multiplexers provide a convenient means for selectively routing information from one circuit to another. The circuit of figure 10-14 allows one of the four signal sources A, B, C, or D to be presented at the output according to the choice of signals applied to the control inputs. Wire the multiplexer circuit of figure 10-14 using the integrated circuits located on the basic gates job board, and connect the data inputs to logic sources SA-SD. The control inputs (a-d) should be connected to SE-SH, and the state of the output can be determined with one of the indicator LEDs. Determine the effect of each of the inputs upon the output for each of the control settings shown in the table, and record the results.

| Inputs | | | | Outputs | |
| a | b | c | d | out | no effect on output |
|---|---|---|---|---|---|
| L | L | L | L | L | A,B,C,D |
| H | L | L | L | A | B,C,D |
| L | H | L | L | ___ | ___ |
| L | L | H | L | ___ | ___ |
| L | L | L | H | ___ | ___ |

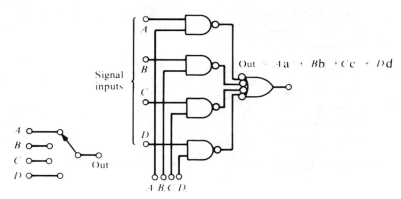

Control inputs  **0.** Closed; **1.** Open

Fig. 10-14

Question 11.  The control inputs of the multiplexer made from basic gates are cumbersome to use.  Describe a circuit that could be added to the simple multiplexer to allow the data channel to be controlled by a two-bit binary number.

An elegant MSI implementation of the multiplexer function is found in the 74151 eight-line to one-line data selector/multiplexer located on the MSI gates job board.  The pin configuration of this IC is illustrated in figure 10-15.

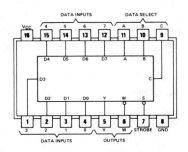

Fig. 10-15

Connect logic source outputs SA, SB and SC to the data select inputs A, B, and C of the 74151 integrated circuit.  Wire eight different TTL signals to the data inputs D0-D7.  (For signals, use other switches, the three outputs of the digital time base, the function generator TTL out, etc).  Note in the table below the identity of the signal connected to each data input.  Finally, connect the strobe input of the 74151 to the MA output of momentary switch MA. Now systematically change the binary data select lines from LLL to HHH and determine which of the input signals appears at the outputs, Y and W.  A convenient method for simultaneously observing the two outputs is to display them on the oscilloscope.

| Data Input | Signal | Data Select | Outputs | |
|---|---|---|---|---|
| | | | Y | W |
| 0 | | LLL | | |
| 1 | | | | |
| 2 | | | | |
| 3 | | | | |
| 4 | | | | |
| 5 | | | | |
| 6 | | | | |
| 7 | | | | |

Depress MA and note the effect on one or more of the signals.

Question 12.   A digital time base is composed of a 1.00 MHz crystal oscillator and 3 decade-dividers. Outputs of 1.00 MHz, 100 kHz, 10 kHz, and 1 kHz are available on 4 separate lines. Design a circuit to multiplex these outputs and allow switch selection of the frequency. Use basic gates in your design.

## 10-4  LATCHES AND REGISTERS

Objectives:  To study the latch or memory function of the cross-coupled inverting gate by observing the latching operation of cross-coupled NAND gates and the technique of gating the inputs to the latch.

Wire the basic NAND gate latch shown in figure 10-16.  Connect Q and Q̄ to indicator lights.  Connect the momentary switch outputs MA and MB to the set (S) and reset (R) inputs.

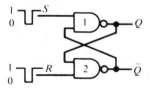

Fig. 10-16

Observe what happens when a momentary 1→0 transition is applied to the gate whose output is LO. Complete the table below.

| Inputs | | Outputs | | | |
|---|---|---|---|---|---|
| | | before pulse | | after pulse | |
| S | R | Q | $\overline{Q}$ | Q | $\overline{Q}$ |
| | H | —— | —— | —— | —— |
| H | | —— | —— | —— | —— |

Do not disconnect this circuit for it will be used as part of the gated latch to be wired next.

Question 13. When the set and reset inputs are in the HI state, what does the basic latch remember?

Question 14. Why should one avoid the application of simultaneous pulses of the same pulse width to the S and R inputs?

Modify the basic latch of figure 10-16 to make the gated latch of figure 10-17.

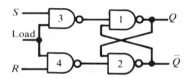

Fig. 10-17

Use SA and SB to provide the S and R inputs and MA to provide the load input. With the load input LO, try all possible combinations of S and R and note the effect on the output. With both S and R inputs LO, apply the load pulse several times and note the effect at Q. Repeat for S = L, R = H and S = H, R = L. Fill in the table below.

| Inputs before load pulse t(n) | | Outputs before load pulse, t(n) | | Outputs after load pulse, t(n+1) | |
|---|---|---|---|---|---|
| R | S | Q | $\overline{Q}$ | Q | $\overline{Q}$ |
| L | L | —— | —— | —— | —— |
| L | H | —— | —— | —— | —— |
| H | L | —— | —— | —— | —— |
| H | H | —— | —— | indeterminate | |

Retain this circuit for the next modification.

Question 15. During what time interval does the latch respond to input data? When are the data latched?

<u>Question 16</u>.  Why does an indeterminate output result at t(n+1) when S and R are both HI?

Change the gated latch to correspond to the data latch circuit illustrated in figure 10-18.  Connect a logic source output to the D input and the function generator (FG) TTL output to the clock input.  Set the FG to 0.5 Hz.  Connect the clock signal and the data latch output to indicators LA and LB.

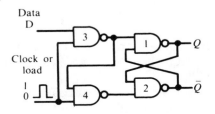

Fig. 10-18

Observe the data transfer times with respect to the clock signal when the D level is changed at various times in the clock cycle.

<u>Results:</u>

If D is changed when clock is HI:  _____

If D is changed when clock is LO:  _____

Disconnect the latch circuit.  NOTE:  the 7475 quad latch IC, studied in Chapter 4, combines 4 data latches of the type studied here in one IC package.


10-5  FLIP-FLOPS

Objectives:  To investigate the response of flip-flops to signals at the data, clock, and preset inputs and to appreciate the differences in the responses of representative flip-flop types by observing the behavior of the 7474 edge-triggered D FF, the 7476 level-triggered JK FF, and the 74112 edge-triggered JK FF.

Install the flip-flop job board.  Locate the 7474 dual D flip-flop. Connect the flip-flop as illustrated in figure 10-19.  Set SA to logic HI and depress MB to clear the flip-flop.

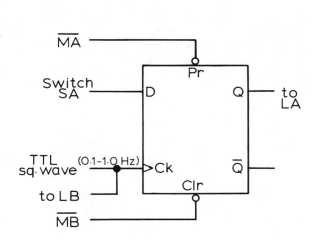

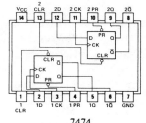

7474
Dual D flip-flop

**FUNCTION TABLE**

| INPUTS | | | | OUTPUTS | |
|---|---|---|---|---|---|
| PRESET | CLEAR | CLOCK | D | Q | $\bar{Q}$ |
| L | H | X | X | H | L |
| H | L | X | X | L | H |
| L | L | X | X | H* | H* |
| H | H | ↑ | H | H | L |
| H | H | ↑ | L | L | H |
| H | H | L | X | $Q_0$ | $\bar{Q}_0$ |

Fig. 10-19

By comparing the time relationships of the clock (LB) to the Q output (LA), note which edge (leading or falling) of the clock causes the information present at the D input to be transferred to the Q output. Set SA to the LO position and repeat the experiment.

> Information at D is transferred to Q on the _____
> edge of the clock.

Now test the time of the response of the flip-flop output to the signal at the preset and clear inputs and note whether these are dependent on the time of the edges of the clock signal.

> A _____ logic level at present causes the Q output
> level to become _____.

> A _____ logic level at clear causes the Q output
> level to become _____.

> Are preset and clear dependent on the clock? _____.

With a very low frequency clock signal, watch for the D input information to be transferred to the Q output. Before the next clock transition, change the information at the D input to the opposite logic level and note whether the new input information can be transferred before the next appropriate clock edge.

> Can the D input information be transferred before the clock edge?
> _____

Now the distinction between edge-triggered and level-triggered flip-flops will be drawn by comparing the 7476 JK flip-flop to the 74112 edge-triggered JK flip-flop. Do not use the 74LS76 in this experiment because it is also an edge-triggered flip-flop.

186

To illustrate this distinction, first wire the 7476 circuit shown in figure 10-20(a). Depress MB to clear the flip-flop. Set the J and K inputs LO, and note the effect of toggling MA. Set J = H and K = H and toggle MA.

On which transition of CK does Q change states? _____

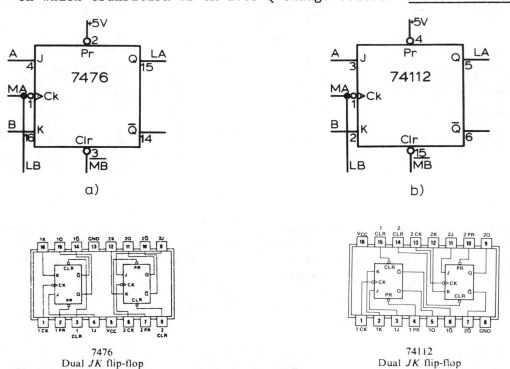

Fig. 10-20

Clear the flip-flop, and depress and hold MA. Bring J to logic HI and back to logic LO without releasing MA. Release MA and note the effect on Q. Note that if J is logic HI at <u>any time</u> while the clock is HI, data will be transferred to the output on the falling edge of the clock.

Disconnect the circuit and wire the 74112 in the same configuration as shown in figure 10-20(b). Repeat the experiment on the loading and transmission of data at the J input.

Describe the differences in the behavior of the 7476 and 74112 flip-flops. _____

_____

<u>Question 17.</u>  Which of the two flip-flops is a master-slave?  Are there any types of flip-flops that combine edge-triggering and master-slave operation?  Of what use are these latter devices?

## 10-6 COUNTERS

Objectives: To study the counting operation and its implementation with flip-flops and MSI counters by connecting flip-flops into binary up and down counters and by using a 74190 decade counting unit IC as a frequency divider.

Wire the four-bit binary counter of figure 10-21 with the two 7476 dual JK flip-flops on the flip-flop job board. Use MA for the count input, $\overline{MB}$ for the clear, and SA for the preset. Set SA to logic HI. Clear the register and advance the count. Fill in the count sequence table.

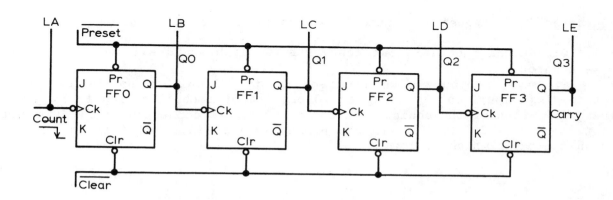

Fig. 10-21

| Count | Q3 | Outputs Q2 | Q1 | Q0 | | Count | Q3 | Outputs Q2 | Q1 | Q0 |
|-------|----|----|----|----|----|-------|----|----|----|----|
| 0 | — | — | — | — | | 8 | — | — | — | — |
| 1 | — | — | — | — | | 9 | — | — | — | — |
| 2 | — | — | — | — | | 10 | — | — | — | — |
| 3 | — | — | — | — | | 11 | — | — | — | — |
| 4 | — | — | — | — | | 12 | — | — | — | — |
| 5 | — | — | — | — | | 13 | — | — | — | — |
| 6 | — | — | — | — | | 14 | — | — | — | — |
| 7 | — | — | — | — | | 15 | — | — | — | — |
| | | | | | | 16,0 | — | — | — | — |

Disconnect $\overline{MA}$ from the Ck input of flip-flop 0 and connect in its place a 1 Hz TTL square wave from the FG. Also connect the FG output to indicator LA. Momentarily depress the clear switch (MB) and observe the counting through one complete cycle.

On which clock transition does the count advance? _____

Question 18. If the change in level at Q of each flip-flop occurs 25 ns after the HI→LO transition at T, when does the LO→HI transition occur relative to the trailing edge of the eighth input pulse?

Question 19. How many binary flip-flops would be required for a counter with a capacity sufficient to count 700 pulses?

Connect Q3 to the B input of the CTFM and the FG TTL output to the A input of the CTFM (leave the FG connected to the Ck input of flip-flop A). Set the CTFM to the frequency ratio mode. Set the FG to 10 kHz and record the frequency ratio f(Ck)/f(Q3). Repeat for Q2, Q1 and Q3.

|    | f(Ck)/f(Q) |
|----|-----------|
| Q0 | _____ |
| Q1 | _____ |
| Q2 | _____ |
| Q3 | _____ |

Question 20. Suggest a practical application for this circuit.

Disconnect the CTFM, but leave the FG, lights and switches connected to the counter.

To make a binary down counter, modify the binary up counter of the preceeding experiment by using the $\overline{Q}$ outputs of each flip-flop to toggle the next flip-flop on the chain. Leave the indicators connected to the Q outputs. Set the FG to $\approx$ 1 Hz, toggle the preset switch (SA), and fill in the count sequence table which follows:

| Count | Outputs | | | | Count | Outputs | | | |
|-------|---------|----|----|----|-------|---------|----|----|----|
|       | Q3 | Q2 | Q1 | Q0 |       | Q3 | Q2 | Q1 | Q0 |
| 0 | __ | __ | __ | __ | 8 | __ | __ | __ | __ |
| 1 | __ | __ | __ | __ | 9 | __ | __ | __ | __ |
| 2 | __ | __ | __ | __ | 10 | __ | __ | __ | __ |
| 3 | __ | __ | __ | __ | 11 | __ | __ | __ | __ |
| 4 | __ | __ | __ | __ | 12 | __ | __ | __ | __ |
| 5 | __ | __ | __ | __ | 13 | __ | __ | __ | __ |
| 6 | __ | __ | __ | __ | 14 | __ | __ | __ | __ |
| 7 | __ | __ | __ | __ | 15 | __ | __ | __ | __ |
|   |    |    |    |    | 16,0 | __ | __ | __ | __ |

What does the counter read after the clear pulse? _____

What does the counter read after the preset pulse? _____

Set the FG to 10 kHz and observe the FG TTL output on Ch 1 of the scope. Use Ch 1 as the scope trigger. Connect Q0 to Ch 2 and observe the waveforms with respect to the clock signal. Repeat for Q1, Q2 and Q3.

Sketch all five waveforms.  Be certain to preserve the time relationships among the waveforms in your sketch.

Disconnect the binary counter.

Next the 74190 decade up/down counter will be studied.  Install the counter job board.  Wire the decade up/down counter and display circuit shown in figure 10-22.

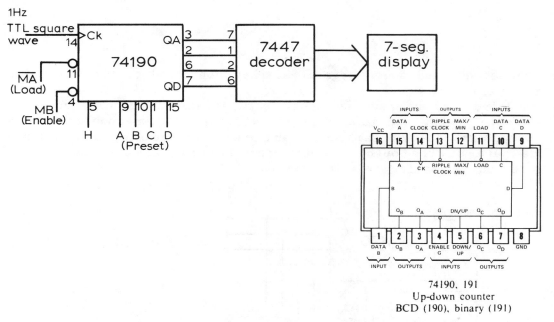

Fig. 10-22

Set the preset switches to HHHH and load the data.  Observe the count and demonstrate that the data can be loaded at any point during the count sequence.  Note the effect of the down/up (D/U) control.  Load several numbers into the counter and observe the results.

Question 21. Note that the 74190 has no external clear input.  Suggest a method to clear the counter.

Does the loading occur synchronously with the clock? _____

Determine the frequency of each of the counter outputs relative to the clock input for several clock frequencies between 10 Hz and 1 M Hz.

| Clock | QA | QB | QC | QD |
|-------|-----|-----|-----|-----|
| ____Hz | ____Hz | ____Hz | ____Hz | ____Hz |
| _____ | _____ | _____ | _____ | _____ |
| _____ | _____ | _____ | _____ | _____ |

Describe your method for measuring the above frequencies.

Leave the decoder and seven segment display connected to the 74190 for the following experiment.

Next the 74190 will be reconnected to form a variable modulus counter. As shown in figure 10-23, the IC counter is wired as a down counter with a gate at the output to detect a zero count and preset the counter to a desired modulus.  The gate at the output of the counter can be constructed from equivalent gates on the basic gates job board.

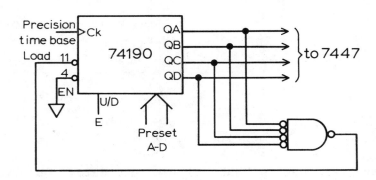

Fig. 10-23

List the IC gates that you plan to use to implement the OR function.

Set the time base to 1 Hz, set the U/D control for down counting, and set the preset switches to LHLH.  Observe and describe the behavior of the counter for this preset value and several others.

Question 22.  Why is this circuit called a variable modulus counter?

Now use the 1 MHz output of the time base as the Ck input to the counter, and measure the frequency of the load pulses as a function ot the magnitude of the number at the preset input.  Complete the table for several preset values.

| Preset Data | f(load) |
|---|---|
| LLLL | _____ |
| ____ | _____ |
| ____ | _____ |
| ____ | _____ |
| HHHH | _____ |

Determine what happens if numbers outside the range 0-9 are applied to the preset input.  Explain.

Question 23.  It is desired to produce a clock with selectable periods of 100 ms, 200 ms, 300 ms, 400 ms and 500 ms.  Design in block diagram form a circuit made from two 74190's that provides these periods.  The preset control should read directly in hundreds of milliseconds.  Do the periods have 50% duty cycles?

10-7 COUNTER GATING

Objectives: To consider the special need of a counter gate not to produce a countable edge at the output due to a change in the control signal by wiring and testing a waveform gate that can be asynchronously opened and closed without introducing counting errors.

The simple counter gate introduced in Chapter 4 has the disadvantage that it allows the control signal to be counted when the count signal is at logic 1. In the following experiment you will construct and investigate the asynchronous waveform gate shown in figure 10-24. Wire the circuit and verify that the control signal allows pulses through the gate only when it is HI. Also show that control signal transitions do not appear at the gate output. Be sure to try both normally HI and normally LO count signals. Record your data below.

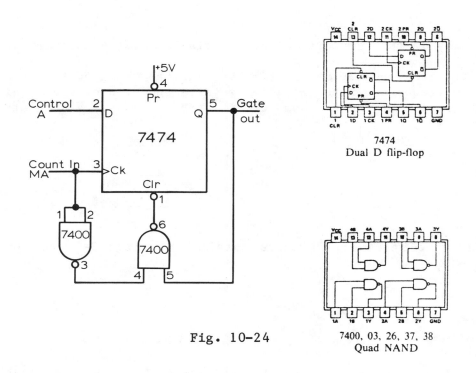

Fig. 10-24

7474
Dual D flip-flop

7400, 03, 26, 37, 38
Quad NAND

Question 24. What constraint is there on the triggering characteristic of the flip-flops of the asynchronous waveform gate?

Question 25. What other components are required to construct a complete counter input gate?

## 10-8  SHIFT REGISTERS

Objectives:  To introduce the shift register as a serial-to-parallel and parallel-to-serial data converter by observing the operation of the shift and load functions of a 4-bit shift register IC.

The 74195, 4-bit parallel access shift register is located on the flip-flop job board.  Connect the circuit as illustrated in figure 10-25a, select the shift mode (logic HI), and clear the register.  Move a HI into the shift register by moving SE to the HI position and depressing MA.

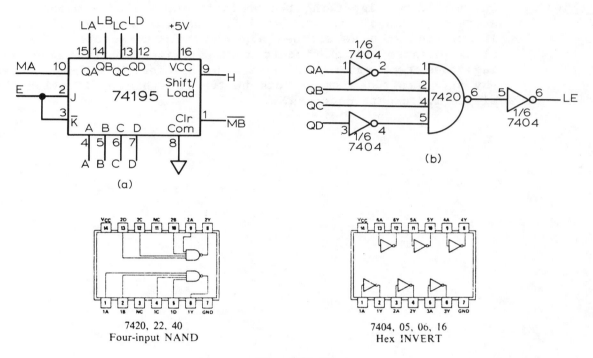

Fig. 10-25

At which output does the serial data appear after the first clock pulse?  _____

Depress MA three more times.

How many clock pulses are required for the first HI entered to appear at QD?  _____

Now the parallel loading feature of the 74195 will be studied.  Select the load mode and enter LLHH from the logic switches by depressing MA.  Return the circuit to the shift mode and clock the data out.

194

<u>Question 26.</u>   Often the BCD digits of a DMM are multiplexed, that is, they are presented to the display one digit at a time at a rate that is undetectable to the human eye.  Devise a method for transmitting such multiplexed BCD data from one instrument to another over a single conductor.

The shift register may be used to sense a given sequence of serial binary bits (at the parallel outputs).  A decoder is all that is necessary to carry out this function.  Add the decoder shown in figure 10-25b to the shift register circuit.  Clear the register, and serially enter binary LHHL.  Note what happens when the complete four bit sequence has been entered.  Try several other four bit sequences.

<u>Question 27.</u>   One method of signalling the beginning or end of a message from a remote source is to send a special pattern of binary information such as encoded alphabetical characters.  Sketch a block diagram of a circuit that utilizes 74195's and 7485 magnitude comparators to detect user-selectable 8-bit codes.  Why are magnitude comparators better for this application than simple logic gate decoders?

# UNIT 11. DIGITAL DEVICES AND SIGNALS

The operating characteristics of common integrated circuits are explored in these experiments. Logic families including TTL, low power Schottky (LS) TTL and CMOS are studied, and open collector and tristate gate types are investigated. Two widely-used IC monostable multivibrators, a simple pulse height discriminator, and an arithmetic logic unit (ALU) are examined.

| Equipment | Parts |
|---|---|
| 1. Basic station | 1. Resistors: 470 Ω, 1 kΩ(2), 4.7 kΩ, 10 kΩ, 22 kΩ, 33 kΩ |
| 2. Basic gates job board | 2. Capacitors: 0.01 µF(3) |
| 3. Op amp job board | 3. LED |
| 4. Utility job board | 4. Scope probe |
| 5. Reference job board | |
| 6. MSI gates job board | |
| 7. Advanced gates job board | |
| 8. Signal conditioning job board | |

## 11-1 LOGIC FAMILIES

Objectives: To investigate the HI and LO logic level voltage ranges for logic gates in three logic families; to determine the gate input current and its direction.

Install the basic gates job board in the frame. Attach the VRS ($\pm$ 10 V output) to one of the 7400 NAND gate inputs as shown in figure 11-1. CAUTION: Keep the VRS voltage in the range of 0 to +5 V. Measure the output voltage as a function of the VRS input voltage. Use the DMM to measure both the input and output voltages. Report your results in the table.

| Input Voltage | Output Voltage |
|---|---|
| _____ | _____ |
| _____ | _____ |
| _____ | _____ |
| _____ | _____ |
| _____ | _____ |
| _____ | _____ |

Question 1. Within what range of VRS voltages is the output solidly HI? Within what range is the output solidly LO? What range of voltage should be avoided? Why?

Now set the VRS to 0 V and insert a current meter between the VRS $\pm$ 10 V output and the NAND gate input. Measure the input current I(iL) for a LO input and note its direction. Repeat for the HI level input current I(iH) by setting the VRS to +5 V and measuring the input current.

I(iL) = _____

I(iH) = _____

In which input state is the source required to absorb current from the gate input? _____

Obtain an advanced gates job board and locate the 74LS00 NAND gate on the board. The pinout is identical to the 7400 NAND gate shown in figure 11-1. Attach the VRS to one input and +5 V to the other as in figure 11-1. Vary the VRS in the range of 0 to +5 V and obtain the input-output characteristics as you did with the 7400 gate.

Fig. 11-1

| Input Voltage | Output Voltage |
|---|---|
| _____ | _____ |
| _____ | _____ |
| _____ | _____ |
| _____ | _____ |
| _____ | _____ |
| _____ | _____ |

Measure the LO and HI level input currents for the 74LS00 gate as you did for the 7400 NAND gate.

I(iL) = _____

I(iH) = _____

Question 2. What differences were noted in the behavior of the TTL and LS gate? Why are LS gates becoming very popular?

Locate the CMOS quad 2-input NOR gate (4001) on the advanced gates job board. Connect the circuit shown in figure 11-2a. CAUTION: Make certain the VRS voltage does not exceed the power supply voltage V(CC) at pin 14. Vary the VRS input voltage over the range of 0 V to +V(CC). Measure the output voltage as a function of VRS input voltage. Report your results in the table.

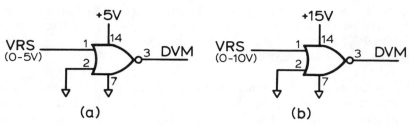

Fig. 11-2

V(CC) = +5 V     **Input Voltage**     **Output Voltage**

Turn the power to the frame OFF.  Disconnect pin 14 on the NOR gate chip from +5 V and instead connect it to +15 V as shown in figure 11-2b.  Turn the frame power back ON and repeat the measurement of the output voltage vs. input voltage for V(CC) = +15 V.  Report your results below.

V(CC) = +15 V     **Input Voltage**     **Output Voltage**

Note:  The CMOS gate input current is much too low to measure with the DMM.

Question 3.  For what range of input voltages was the CMOS gate output soldily HI when operated from +5 V?  From +15 V?  For what ranges was the output LO?  What voltage range should be avoided for operation at +5 V?  For +15 V?

Question 4.  What are the advantages in operating CMOS gates from higher supply voltages (+12 to +15 V)?  What are the disadvantages?

11-2 OPEN COLLECTOR GATES

Objectives: To investigate the operating characteristics of TTL open-collector gates by observing the output resistance of a typical gate as a function of the input logic state, by connecting a gate to drive a LED, by combining open collector gates to produce "wired" logic, and by measuring switching speeds for open collector configurations.

Locate the 7406 open collector driver IC on the advanced gates job board. Refer to the pinout of the IC shown in figure 11-3 and wire the input of one of the gates to SA.  Connect the DMM between the gate output and ground, and measure the resistance of the output transistor for each of the input states. (Do not connect the 1 kΩ pullup resistor, yet).

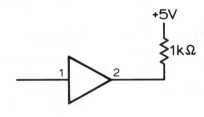

Fig. 11-3

Complete the table below and explain how the driver IC works.

| input state | output R | state of output transistor (on/off) |
|---|---|---|
| LO | _____ | _____ |
| HI | _____ | _____ |

Explanation:




Now change the DMM to the DC volts mode, and connect a 1 k  pullup resistor between the output of the gate and +5 V as shown in figure 11-3. Measure the output voltage of the gate for each of the input states and record the results below.

| input state | output voltage |
|---|---|
| LO | _____ |
| HI | _____ |

The driver IC may be used to control an LED as shown in figure 11-4. Obtain an LED and wire the circuit.  (The flat side (short lead) is the cathode).  Determine the state of the LED for each of the input states.

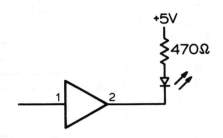

Fig. 11-4

| input state | LED state |
|-------------|-----------|
| LO | _____ |
| HI | _____ |

<u>Question 5</u>.  Explain how the LED is driven by the IC.  Why is the choice of
the value of the resistor critical for proper operation of the
LED?

The outputs of open collector gates may be wired to produce the "wired
logic" effect.  Connect the circuit of figure 11-5 with three of the gates on
the 7407 IC.  Use switches SA, SB, and SC and lamp LA to determine the logic
function performed by the circuit.

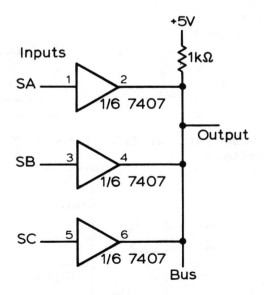

Fig. 11-5

| Inputs | | | |
|---|---|---|---|
| A | B | C | Output, M |
| L | L | L | _____ |
| L | L | H | _____ |
| L | H | L | _____ |
| L | H | H | _____ |
| H | L | L | _____ |
| H | L | H | _____ |
| H | H | L | _____ |
| H | H | H | _____ |

What logic function is performed by the "wired logic" for HI true signals? _____.

What logic function is performed for LO true signals? _____

Connect a 1kΩ pullup resistor to the output of an open collector NAND gate of the 7403 IC on the advanced gates job board, and verify the function table as shown in figure 11-6.

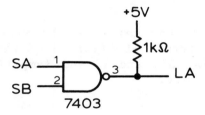

Fig. 11-6

| Inputs | | |
|---|---|---|
| A | B | Output, M |
| LO | LO | _____ |
| LO | HI | _____ |
| HI | LO | _____ |
| HI | HI | _____ |

Now wire two 7403 NAND gates as illustrated in figure 11-7 and determine the resulting logic function.

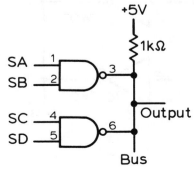

Fig. 11-7

|                | Inputs |        |        |           |
| -------------- | ------ | ------ | ------ | --------- |
| <u>A</u> | <u>B</u> | <u>C</u> | <u>D</u> | <u>Output, M</u> |
| L | L | L | L | _____ |
| L | L | L | H | _____ |
| L | L | H | H | _____ |
| L | H | L | H | _____ |
| L | H | H | H | _____ |
| H | H | H | H | _____ |

Logic function for HI true signals _____

Logic function for LO true signals _____

Next the switching speed of open collector gates will be explored using the circuit of figure 11-8. Assemble the circuit as shown with R = 1 kΩ.

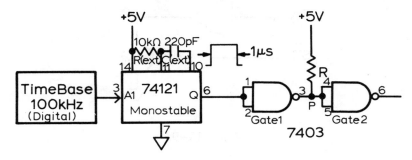

Fig. 11-8

Apply a 100 kHz TTL square wave to one of the inputs of the 74121 monostable multivibrator on the signal conditioning job board. Use a properly adjusted scope probe (see your scope manual), and measure the rise time and fall time (10% to 90%) of the ∼ 1 μs pulse at point P. The load on gate 1 consists of the probe and both inputs of gate 2. Repeat the measurement for the two other pullup resistors listed in the table and record your results.

| <u>R</u> | <u>rise time (10%-90%)</u> | <u>fall time (90%-10%)</u> |
| ------ | ------------------- | ------------------- |
| 470 Ω | _____ | _____ |
| 1 kΩ | _____ | _____ |
| 10 kΩ | _____ | _____ |

<u>Question 6</u>. Open collector logic has often been used in the construction of computer buses, e.g., many different signals may be connected to the same signal line to provide bidirectional communication among different circuits. Sketch a simple open collector bus circuit with two inputs and two outputs.

<u>Question 7</u>. How could the possibility that two different signal sources might assert a LO state on the bus be averted?

Disconnect the circuit, but leave the monostable setup to produce a 1-2 μs pulse.

11-3   TRISTATE GATES

Objectives:   To examine the properties of tristate logic by determining the
              effect of the enable input, by connecting several tristate gates
              in a bus configuration, and by measuring the switching speeds of
              a typical gate.

    The circuit of figure 11-9 may be used to investigate the behavior of
tristate logic.  Each gate on the 74125 IC shown in the figure has an enable
input which is used to control whether or not the gate is in the high
impedance (passive) mode.  Find the 74125 IC on the advanced gates job board
and wire the circuit shown in the figure.

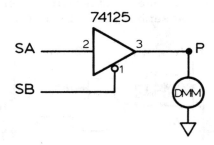

Fig. 11-9

Determine the output voltage of the gate for all four possible combinations of
enable and data inputs, and record your results.

| data input | enable input | Output Voltage |
|------------|--------------|----------------|
| HI | LO | _____ |
| HI | LO | _____ |
| LO | HI | _____ |
| HI | HI | _____ |

In order to determine which output states are asserted it is necessary to tie
the output of the gate to the logic level opposite that which is expected.
Connect a 4.7 kΩ resistor between the output (point P) and +5 V.  Observe the
effect of enabling the gate with the input LO.  In the same way, tie P to
common through the resistor and determine the logic level of the enable input
that asserts a HI at the output.

| data input | enable input | resistor tied to | output voltage |
|------------|--------------|------------------|----------------|
| LO | LO | +5V | _____ |
| LO | HI | +5V | _____ |
| HI | LO | COM | _____ |
| HI | HI | COM | _____ |

    What logic level enables the output of the tristate gate?
_____

Question 8. Explain the behavior of tristate logic in terms of the internal circuitry of the logic family.

An application of the 74125 IC in a bus configuration is illustrated in figure 11-10. Assemble the circuit. Exercise the circuit by enabling each of the gates individually and by verifying the transfer of data from the input to the output of the gate. Satisfy yourself that inputs that are not enabled have no effect on the bus. Record your results in the table below.

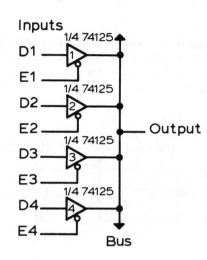

Fig. 11-10

| D1 | E1 | D2 | E2 | D3 | E3 | D4 | E4 | Output |
|----|----|----|----|----|----|----|----|--------|
| —— | —— | —— | —— | —— | —— | —— | —— | —— |
| —— | —— | —— | —— | —— | —— | —— | —— | —— |
| —— | —— | —— | —— | —— | —— | —— | —— | —— |
| —— | —— | —— | —— | —— | —— | —— | —— | —— |

Now deliberately enable various combinations of more than one gate as shown in the table below, and measure the resulting output voltage with the DMM.

| | states of gates | | | output |
|----|----|----|----|--------|
| 1 | 2 | 3 | 4 | voltage |
| HI | LO | DIS | DIS | _____ |
| HI | LO | LO | DIS | _____ |
| HI | HI | LO | DIS | _____ |
| HI | HI | HI | LO | _____ |
| HI | HI | LO | LO | _____ |
| HI | LO | LO | LO | _____ |

204

Explain what happens when more than one gate output is enabled.

As in the case of open collector logic, the speed of tristate logic can be easily measured. Connect the circuit of figure 11-11, and measure the rise and fall times of the tristate gate in the bus configuration. The 7403 NAND gate input serves to load the bus.

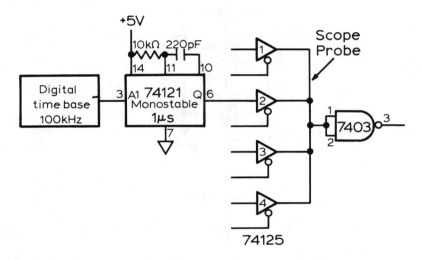

Fig. 11-11

rise time _____          fall time _____

Compare the speeds of open collector and tristate logic gates.

11-4  LEVEL CONDITIONING

Objectives:  To explore two types of level conditioning by wiring and testing
the characteristics of a Schmitt trigger gate and by constructing
a contact bounce eliminator.

Locate the 7414 hex Schmitt trigger IC on the advanced gates job board.
Set the VRS $\pm$ 10 V variable output to 0.00 V, and connect it and the DMM to
one of the Schmitt trigger inverting gates on the IC as shown in figure 11-12.
Observe the state of the output as you carefully increase the voltage.  Record
the voltage at which the output makes a HI $\rightarrow$ LO transition.  Now decrease the
voltage and repeat the measurement for a LO $\rightarrow$ HI transition at the output.
Repeat each measurement several times, and record the results below.

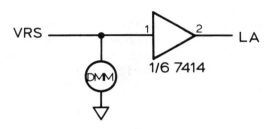

Fig. 11-12

Input Voltage

| HI $\rightarrow$ LO | LO $\rightarrow$ HI |
|---|---|
| _____ | _____ |
| _____ | _____ |
| _____ | _____ |
| _____ | _____ |

mean _____   _____

standard deviation _____   _____

Question 9.  Describe the behavior of the voltage input at the trigger point.

Question 10.  Compare the behavior of the 7414 gate with that of a standard
TTL gate as determined in experiment 11-1.

Disconnect the DMM and VRS from the gate.  Connect the FG output to Ch 1
of the scope and adjust the FG output to give a 0-4 V p-p, 10 kHz sine wave.
Connect this signal to the Schmitt trigger input.  Observe the output of the
gate on Ch 2 while triggering on Ch 1.  Observe the leading and falling edges
of the output for various frequencies down to about 100 Hz.  Sketch the
observed waveform in the following space.

206

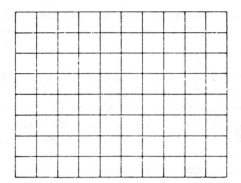
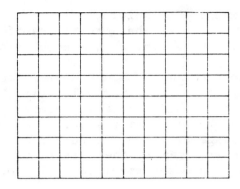

<u>Question 11</u>.  Why is the Schmitt trigger gate useful for observing the state of slowly changing signals?

In order to provide a convenient means of observing the effects of debouncing a switch, an SPDT switch will be simulated by the relay investigated in Unit 6.  Connect the relay in the configuration illustrated in figure 11-13 using a 7400 IC on the basic gates job board.  Set the FG for 50 Hz, and observe the debounced and raw signals on Ch 1 and Ch 2, respectively.  Sketch the waveforms below.

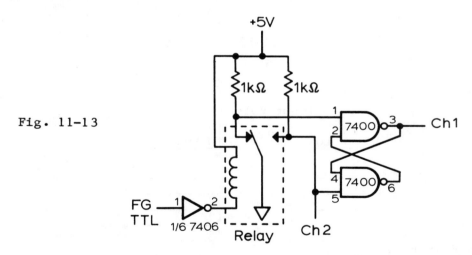

Fig. 11-13

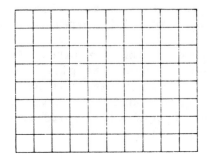

Measure the ratio of the frequencies of the two signals to that of the FG frequency using the CTFM.

f(raw)/f(in)          f(debounced)/f(in)

_____         _____

Question 12. Explain the difference in the two frequencies in terms of the operation of the cross coupled NAND gate debouncer circuit.

## 11-5  IC MONOSTABLES

Objectives: To investigate the operational characterisitcs of monostable multivibrators by determining the dependence of the pulse width upon the passive components and by observing the effects of retriggering upon a retriggerable and non-retriggerable monostable multivibrator.

Use the 74121 monostable multivibrator on the signal conditioning job board, and wire the circuit of figure 11-14 with R = 10 kΩ and C = 0.01 μF. Power up the circuit, turn on the FG, and adjust the frequency to about 5 kHz. Set the scope trigger for Ch 2, and switch SA to the LO state. Record the waveform in the space provided.

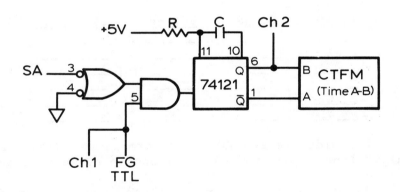

Fig. 11-14

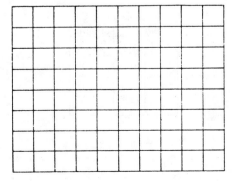

What is the effect of switching SA to HI?

Gradually increase the frequency of the FG until the period of the TTL square wave is shorter than the pulse width. Describe the effect.

Is the 74121 retriggerable or non-retriggerable? _____

Question 13. Of what practical use is the ability to retrigger a monostable multivibrator?

Now determine the effect of changing R and C. This may be accomplished by placing one and two 0.01 μF capacitors in parallel with C and by changing R to 22 kΩ and 33 kΩ. Measure the pulse widths resulting from several combinations of R and C and record the results below. A simple way to measure the pulse width is to connect the Q and Q̄ outputs of the 74121 to the A and B inputs of the CTFM, operated in the Time A-B mode, as shown in figure 11-14.

| R, kΩ | C, μF | RC, μs | t(pw), μs |
|-------|-------|--------|-----------|
| _____ | _____ | _____ | _____ |
| _____ | _____ | _____ | _____ |
| _____ | _____ | _____ | _____ |
| _____ | _____ | _____ | _____ |
| _____ | _____ | _____ | _____ |

Graph 1. Plot pulse width versus RC and determine the slope of the resulting straight line. What is the relationship between RC and t(pw)?

Find the 74123 dual monostable on the signal conditioning job board. The pinout of this popular IC is illustrated along with its timing diagram and function table in figure 11-15.

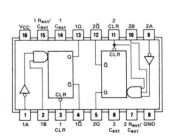

**FUNCTION TABLE**

| INPUTS | | | OUTPUTS | |
|--------|---|---|---------|---|
| CLEAR | A | B | Q | Q̄ |
| L | X | X | L | H |
| X | H | X | L | H |
| X | X | L | L | H |
| H | L | ↑ | ⊓ | ⊔ |
| H | ↓ | H | ⊓ | ⊔ |
| ↑ | L | H | ⊓ | ⊔ |

Fig. 11-15

TYPICAL OUTPUT PULSE WIDTH
vs
EXTERNAL TIMING CAPACITANCE

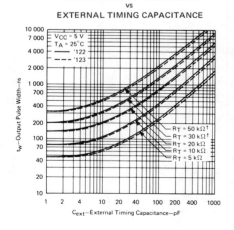

The timing diagram may be used for C(ext) < 1000 pF. For C(ext) > 1000 pF, the pulse width, t(pw), in nanoseconds is given by the following equation:

$$t(pw) = 0.32 \ R(ext) \ C(ext) \ (1 + \frac{0.7}{R(ext)}).$$

Select R(ext) = 10 kΩ and C(ext) = 0.01 μF, and connect the FG TTL out to pin 1 of the IC. Measure t(pw) (scope or CTFM?) and compare with the value calculated from the equation.

t(pw) measured    = _____

t(pw) calculated = _____

Is the measured value correct to within the component tolerances? _____

Now connect pin 2 of the IC to common and note the effect.

Disconnect the inputs of the monostable, wire the FG to pin 2, and note the result.

What must be done to make the one-shot work?

Finally, perform an experiment to determine whether the 74123 is retriggerable or not. Describe the experiment and note the results

Question 14.    Suppose that a stream of equally spaced bubbles is required to pass through a narrow tube in an automated chemical instrument. Design a simple circuit using a 74123 and any other necessary components to light an LED if the stream is not carrying bubbles spaced a given time interval apart.

<u>Question 15.</u>  Design a circuit that will produce a 1 μs negative pulse exactly 6 μs after the occurence of the falling edge of a certain TTL signal.


## 11-6  WINDOW DISCRIMINATOR

Objective:  To investigate the properties of a pulse height window discriminator by wiring the circuit and determining the upper and lower threshold voltages.

Wire the circuit of figure 11-16.  Use two LM311 comparators, one on the op amp job board and the other on the signal conditioning job board, and a 7486 exclusive-OR gate on the basic gates job board.

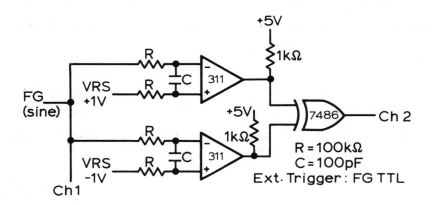

Fig. 11-16

Adjust the FG sine wave output to produce a 5 V p-p, 5 kHz signal symmetrical about 0 V.  Sketch the input and output waveforms below, and indicate the reference voltages on the input waveform.

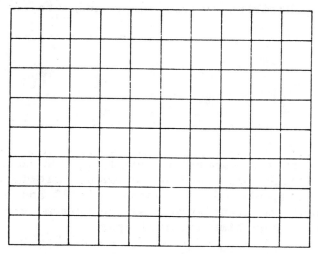

Gradually adjust the offset of the FG until the peaks of the sine wave fall below 1 V, and sketch the resulting waveforms below.

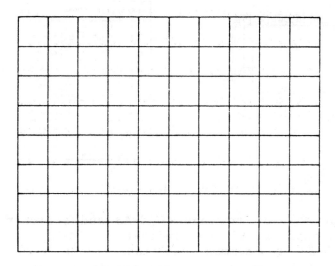

By carefully adjusting the offset of the FG, determine the upper and lower threshold voltages for the appearance of a transition at the output of the circuit.

upper threshold _____

lower threshold _____

Question 16. Circuits similar to the one above may be used in photon counting to discriminate between single photons, shot noise, and multiple photon events. Describe in detail how such a arrangement might be used to count single photon events.

## 11-7  ARITHMETIC LOGIC UNIT

Objectives: To investigate the truth table of an arithmetic logic unit (ALU); to observe the logic functions and arithmetic functions that can be performed.

Locate the 74181 arithmetic logic unit on the MSI gates job board. Note that it is a 24 pin IC and is wider than the normal 14 and 16 pin DIPs. Because of the width, the chip covers two inner rows of connectors on the job board. Note the pinout and function table given in figure 11-17.

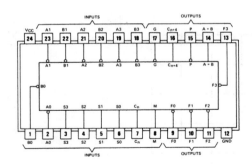

| SELECTION | ACTIVE-HIGH DATA | | |
|---|---|---|---|
| | M = H | M = L; ARITHMETIC OPERATIONS | |
| S3 S2 S1 S0 | LOGIC FUNCTIONS | $C_n$ = H (no carry) | $C_n$ = L (with carry) |
| L L L L | F = $\overline{A}$ | F = A | F = A PLUS 1 |
| L L L H | F = $\overline{A + B}$ | F = A + B | F = (A + B) PLUS 1 |
| L L H L | F = $\overline{A}$B | F = A + $\overline{B}$ | F = (A + $\overline{B}$) PLUS 1 |
| L L H H | F = 0 | F = MINUS 1 (2's COMPL) | F = ZERO |
| L H L L | F = $\overline{AB}$ | F = A PLUS A$\overline{B}$ | F = A PLUS A$\overline{B}$ PLUS 1 |
| L H L H | F = $\overline{B}$ | F = (A + B) PLUS A$\overline{B}$ | F = (A + B) PLUS A$\overline{B}$ PLUS 1 |
| L H H L | F = A $\oplus$ B | F = A MINUS B MINUS 1 | F = A MINUS B |
| L H H H | F = A$\overline{B}$ | F = A$\overline{B}$ MINUS 1 | F = A$\overline{B}$ |
| H L L L | F = $\overline{A}$ + B | F = A PLUS AB | F = A PLUS AB PLUS 1 |
| H L L H | F = $\overline{A \oplus B}$ | F = A PLUS B | F = A PLUS B PLUS 1 |
| H L H L | F = B | F = (A + $\overline{B}$) PLUS AB | F = (A + $\overline{B}$) PLUS AB PLUS 1 |
| H L H H | F = AB | F = AB MINUS 1 | F = AB |
| H H L L | F = 1 | F = A PLUS A* | F = A PLUS A PLUS 1 |
| H H L H | F = A + $\overline{B}$ | F = (A + B) PLUS A | F = (A + B) PLUS A PLUS 1 |
| H H H L | F = A + B | F = (A + $\overline{B}$) PLUS A | F = (A + $\overline{B}$) PLUS A PLUS 1 |
| H H H H | F = A | F = A MINUS 1 | F = A |

Fig. 11-17

Connect switch outputs SA-SD to inputs A0-A3 and switches SE-SH to inputs B0-B3. Connect outputs F0-F3 to indicators LA-LD. Initially connect selection inputs S0-S3 to common and mode input M=HI to select logic functions. Verify that F=$\overline{A}$ and record your results below.

| | Select | Inputs | | Outputs |
|---|---|---|---|---|
| M | S3 S2 S1 S0 | A3 A2 A1 A0 | B3 B2 B1 B0 | F3 F2 F1 F0 |
| $\overline{H}$ | L L L L | L L L L | X X X X | __ __ __ __ |
| | | L L L H | | __ __ __ __ |
| | | L H L H | | __ __ __ __ |
| | | H L H L | | __ __ __ __ |

Now select other logic functions by changing the logic levels at the select (S3 S2 S1 S0) inputs. Verify the function table for several of the logic functions shown in figure 11-17.

| | Select | Inputs | | Outputs |
|---|---|---|---|---|
| M | S3 S2 S1 S0 | A3 A2 A1 A0 | B3 B2 B1 B0 | F3 F2 F1 F0 |
| $\overline{H}$ | __ __ __ __ | __ __ __ __ | __ __ __ __ | __ __ __ __ |
| | __ __ __ __ | __ __ __ __ | __ __ __ __ | __ __ __ __ |
| | __ __ __ __ | __ __ __ __ | __ __ __ __ | __ __ __ __ |
| | __ __ __ __ | __ __ __ __ | __ __ __ __ | __ __ __ __ |
| | __ __ __ __ | __ __ __ __ | __ __ __ __ | __ __ __ __ |
| | __ __ __ __ | __ __ __ __ | __ __ __ __ | __ __ __ __ |
| | __ __ __ __ | __ __ __ __ | __ __ __ __ | __ __ __ __ |

Now change the mode M to LO to select arithmetic functions. Connect the carry in, Cn, to logic HI (no carry) and the carry out, C(n+4), to indicator LE. Verify several of the arithmetic functions shown in figure 11-17.

|  |  | Select | | | | Inputs | | | | | | | | Outputs | | | | |
|---|---|---|---|---|---|---|---|---|---|---|---|---|---|---|---|---|---|---|
| M | Cn | S3 | S2 | S1 | S0 | A3 | A2 | A1 | A0 | B3 | B2 | B1 | B0 | F3 | F2 | F1 | F0 | C(n+4) |
| L | H | — | — | — | — | — | — | — | — | — | — | — | — | — | — | — | — | — |
|  |  | — | — | — | — | — | — | — | — | — | — | — | — | — | — | — | — | — |
|  |  | — | — | — | — | — | — | — | — | — | — | — | — | — | — | — | — | — |
|  |  | — | — | — | — | — | — | — | — | — | — | — | — | — | — | — | — | — |
|  |  | — | — | — | — | — | — | — | — | — | — | — | — | — | — | — | — | — |
|  |  | — | — | — | — | — | — | — | — | — | — | — | — | — | — | — | — | — |

Change the carry in to LO (with carry) and note its effect for several functions.

Question 17. What ALU output (F3-F0) would you predict for the following input conditions? Explain.

| S3 | S2 | S1 | S0 | M | Cn | A3 | A2 | A1 | A0 | B3 | B2 | B1 | B0 |
|---|---|---|---|---|---|---|---|---|---|---|---|---|---|
| H | L | H | L | L | L | L | H | H | L | L | L | H | H |

# UNIT 12. INTRODUCTION TO MICROCOMPUTERS

In this unit, the AIM-65 microcomputer system is introduced. The system is programmed in the BASIC language and in assembler language. The methods used by the BASIC interpreter to store integer and floating point variables are demonstrated. Linkages between BASIC and assembler language routines are developed. The subroutine concept is used both in BASIC and in assembler. Finally, a simple example of an interrupt driven system is given. The goal of these experiments is to acquaint the user with the software system that resides in the BASIC equipped AIM-65. Subsequent experiments in Units 13 and 14 explore the hardware aspects of the AIM-65.

It is not intended that these experiments be an exhaustive study in computer programming either in BASIC or in machine language. Instead, the techniques needed for experiments in Units 13 and 14 are introduced. Further study of BASIC and the machine language monitor should continue with the AIM-65 User's Guide and the AIM-65 BASIC Language Guide.

This experiment has no blanks to fill in. All the observations you will make can be recorded on the computer printout. Therefore, the report for this experiment will consist of sections of the computer printout, labeled and attached to the answers and explanations requested in the experiment.

|  Equipment  |  Parts  |
|---|---|

1. Basic station
2. AIM-65 Microcomputer with
   4K RAM and BASIC chip set


## 12-1  INITIALIZATION OF BASIC

Consult the instructor for directions for turning the AIM-65 system on. Upon turning the power on, the system will print:

ROCKWELL AIM-65

and the character display will show a "<" at the left-most position. At this point the AIM-65 is in the monitor mode. For the moment, this mode will be ignored, and the BASIC interpreter will be used instead. To enter the interpreter, depress the 5 key. The display will show:

MEMORY SIZE?

Depress the RETURN key. This causes the interpreter to reserve all the memory available for the use of the BASIC system. The printer will print the previous question, and the display will show:

215

WIDTH?

The answer to this question tells the BASIC interpreter how many columns the output device has.  If RETURN is depressed the computer uses a default value of 20, which is appropriate for the 20 column display and printer of the AIM-65.  At this point, the printer prints the last question and then prints the number of memory locations available and the BASIC title.  The display at this point shows the last line printed, except that the cursor character, ^ , is displayed in column one.  As soon as anything is typed on the keyboard, the old line disappears from the display.  The cursor shows where the next character typed into the computer will go.

The twenty column width of the display and the printer causes some difficulty in BASIC.  BASIC statements are allowed to be sixty characters long.  As the statement is typed, the twenty-first character input causes the first twenty characters to be printed on the printer.  Similarly, the forty-first character causes the second twenty characters to be printed.  When RETURN is typed the remaining characters are printed.  The display scrolls right to left as more than twenty characters are entered.  If a mistake is made while typing, the DEL key backspaces and deletes one character at a time.

During your experiments you may accidentally press keys which cause the BASIC interpreter program to become lost or switched off.  The symptoms of this condition are a nonsense display or an unresponsive keyboard.  BASIC can be reinitialized (restored) by pressing the RESET button on the main board or by turning off the power for a minute or so.  The computer comes back in monitor, so the "5" key should be pressed to reenter BASIC.

## 12-2  DIRECT AND INDIRECT MODES

The BASIC interpreter has two modes of operation, called DIRECT mode and INDIRECT mode.  In direct mode, the command given to the BASIC interpreter is executed immediately.  This is the first mode that will be used to introduce the BASIC commands.

A.  The PRINT Statement

The print command is the standard method by which BASIC programs communicate with the outside world.  Try the following.  Be sure to end each line with a RETURN.

PRINT "HELLO"

The "HELLO", including the quotes, is a CHARACTER STRING.  Character strings are useful as labels, or in printing messages.  They are not, however, useful in calculations.

The paper can be advanced one line at a time by pressing the LF key.  Do not pull the paper through for viewing or tear-off.

Numbers can also be printed.  Try the following, and attach the printer output that results.

```
PRINT 2
PRINT 2/9
PRINT .123456789
```

```
PRINT 12345.678912345
PRINT 6/3*4-2
```

From the previous examples, it is obvious that BASIC does not always return the same number that was input. This is because the method BASIC uses to store numbers is limited to approximately 9 decimal digits. Experimentally determine the number of digits BASIC appears to retain, and how the extra digits are rounded off.

Question 1. How many digits in a number does BASIC retain and how is the number rounded off? Label the section of relevant printout P1 and attach it to your answer. Explain how your answer was obtained from the experiment recorded in the printout.

Another method of expressing numbers in BASIC is the so-called exponential or scientific notation. Try, for example:

```
PRINT 1.0E2
PRINT 1.345E-2
PRINT 1234.5E18
```

BASIC selects between normal and scientific output depending upon the magnitude of the number. Experimentally determine the change-over points, to the nearest power of ten.

Question 2. Above what value does BASIC use exponential form for a number? Explain how your experiment demonstrates your answer and attach the relevant printout (labeled P2).

There is a limit to the magnitude of numbers allowed in BASIC, even in exponential form, Enter, for example: PRINT 1E132. The response of the computer is one of several ERROR MESSAGES that the computer uses to indicate an operation that is illegal in BASIC or impossible. The OV "message indicates "overflow", i.e., too large a number. Try, PRINT 56/0. Experimentally determine, to the nearest power of 10, the maximum magnitude a digit can have in BASIC. There is a similar limit on how small a non-zero number can be. Determine the maximum magnitude of a negative exponent in BASIC and what the response of the computer is when the maximum is exceeded.

Question 3. What are the maximum values for positive and negative exponents for numbers in BASIC? What happens when these limits are exceeded? Attach and explain the printout (labeled P3) that demonstrates your answers.

## B.  ASSIGNMENT Statements

The primary utility of the computer in the laboratory is its ability to store and manipulate numerical quantities.  In BASIC, numerical values are stored as VARIABLES.  Each variable is given a name, by which BASIC identifies the area of memory wherein the associated numerical value is stored.  In AIM-65 BASIC, the variable name consists of two characters, the first of which must be alphabetic.  The first time a variable is referenced, BASIC sets aside a memory location to store the numerical value associated with the variable. This occurs both in direct and indirect modes of operation.  The value associated with the variable can be printed, using the PRINT statement.  Try the following:

```
A=3
B=4
PRINT A,B,A*B,A+B,A/B,A-B
```

The arithmetic operators have their usual meanings.  Note that * is used for multiplication.  The F3 key produces the ^ symbol which is used for exponentiation.

A number of arithmetic functions are available in AIM-65 BASIC.  Consult the AIM-65 BK BASIC Reference Card for details.

The manipulation of numbers can lead to some fairly strange looking commands.  Type in:

```
A=A+1
PRINT A
```

The statement "A=A+1" is an assignment statement, and should not be confused with an equality.  It means, change the value of the variable A to its previous value plus one.

If the computer printed 4, it added 1 to the value of A assigned in the previous experiment.  Now both A and B have values of 4.  (Print A*B, for example, to prove this.)  If the computer printed 1 in response to the PRINT A command, the previous value of A had been lost through power-down or reset and was arbitrarily assigned a value of zero when the BASIC interpreter was restarted.

## C.  Indirect Mode

Executing commands in direct mode is useful for demonstrations and for single calculations.  However, in order to use a computer efficiently, we must be able to give the computer a list of instructions to perform, and then turn it loose to perform them.  The list of instructions is the PROGRAM; the creation of the list is PROGRAMMING.  In BASIC, the program consists of a list of commands, each with a statement number.  Enter the following:

```
NEW
10 A=3
20 B=4
30 PRINT A,B,A/B
40 END
```

The NEW command resets BASIC, deleting any program or variable that has been set up in memory. It should be used with caution. Notice that the statements printed out on the printer as they were entered, but that the values of A, B, and A/B were not printed. This is because the program has not yet been RUN. It has, however, been stored by the interpreter. To recall the stored program, type:

    LIST

Now to execute the program, type:

    RUN

Statements can be added to the program by simply entering them. For example, type the following:

        15 PRINT A^3
        37 PRINT C/A
        35 C=5
        LIST

Note from the listing that the new statements have been added in the proper numerical order, regardless of the order in which they were typed. Run the modified program.

To change the quantity calculated in line 37, type:

        37 PRINT A/C
        LIST

Run this program. This method of line replacement can be used to delete program lines. Simply enter the statement number and press RETURN. Do this to statement 37, and verify the result.

The reason for incrementing by ten or so between statement numbers is to allow additional statements to be added between existing ones. Modify the program just listed to print out the six possible ratios of the variables A, B, and C taken two at a time. List and run the program.

Question 4. Attach the listing and execution of your modified program to your report (P4). Comment on your choice of method for modification.

## 12-3 INPUT STATEMENTS

The ability to do computations and to print results is enhanced if the numbers can be input when the program is run, instead of being embedded in the program itself. To do this, the INPUT statement is used. Type the following:

```
NEW
10 INPUT A
20 INPUT B
30 C=A+B
40 PRINT A<B<C
50 END
```

Run this program. Immediately a ? appears on the display. Enter the value you wish to assign to A, and press RETURN. The printer prints the value, and then the ? is again displayed. This time, enter the value you wish to assign to B, again followed by RETURN.

The problem here is the question mark. It is difficult to decide what the computer wants. A modification of the INPUT statement will solve the dilemma:

```
10 INPUT "A=";A
20 INPUT "B=";B
```

Now run the program, and observe the difference.

The above program needs to be restarted every time you want to enter a new set of values for A and B. The program would restart automatically if the END command were replaced by:

```
50 GOTO 10
```

The GO TO command causes the program control to jump to line 10. This creates an endless loop. Run this program through a few cycles. The problem here is how to stop the program -- how to get out of the loop. To break the loop while it is waiting for an input statement, type Control C, i.e., while the control key is depressed, type C, then RETURN.

Some loops involve just continous execution. Such a loop can be created in the above program by putting a "GO TO 30" at line 50. Before you try it, locate key F1 - holding it down is the only way to interrupt this program (except turning off power or resetting). Now try it, but don't waste too much paper.

Question 5. Write a program that will repeatedly sum A and B and print the result starting with A = 10 and B equals 20 and then increase the value of A by 5 and B by 10 on each repetition. Do not run this program unless you are buying the paper. What might eventually halt this program? When?

## 12-4  ARRAYS AND THE DIM STATEMENT

Frequently in scientific uses of the computer, spectra or time sequences are to be processed. It is convenient to store such data in ARRAYS, which are sets of variables under one name which are accessed by number. The BASIC statement that creates an array is the DIM statement. Type the following direct mode commands:

```
          NEW
          DIM A(10)
          A(0)=1
          A(1)=2
          A(10)=10
```

The DIM statement creates an eleven element array [A(0) through A(10)] called
A.  The subsequent statements assign values to the individual array elements.
The number in parentheses is called the SUBSCRIPT.  The subscript can be a
number, a variable, or an expression.  For example try:

```
          PRINT A(0),A(1)
          I=1
          J=10
          PRINT A(I),A(J)
          PRINT A(I-1)
```

## 12-5  FOR - NEXT LOOPS

Very often each element of an array enters into a calculation in the same
way as all other elements of the array.  For example, a constant may be
subtracted from each element of the array:

```
          A(1)=A(1) - 2
          A(2)=A(2) - 2
             .
             .                           Do not type
             .
          A(10)=A(10)-2
```

Clearly, this would create a long program.  A more compact method uses the FOR
- NEXT loop.  Type in the following:

```
          NEW
          10 DIM A(10)
          20 FOR I=0 TO 10
          30 A(I)=I+2
          40 NEXT I
          50 FOR I=10 TO 0 STEP -1
          60 PRINT I,A(I)
          70 NEXT I
          80 END
          RUN
```

The variable I in statement 20 is called the INDEX variable.  It is initially
assigned the value of 0, and the subsequent statements are executed.  At
statement 40, the value of I is assigned its next value, which is one more
than the previous value, and control is transferred back to the command
following the FOR command in statement 20.  This continues until the value
assigned to I becomes greater than the LIMIT value, which in this case is 10.
Program statements 10-40 thus load the array A with values two greater than
the subscript of each element.

The loop in statements 50-70 is similar except that here an explicit increment of -1 is specified. Such a negative increment also reverses the sense of the test at the end of the loop. Now control will be transferred back to the loop until I is less than the limit value. This part of the program prints the values of the subscript and the array element contends in reverse order from A(10) and A(0).

Demonstrate your understanding of these concepts by writing a program that assigns to each element of an eight element array twice the value of the element's subscript. In a separate loop in the same program, print the elements whose subscript is odd in increasing order. Be sure to execute a NEW statement before you begin to enter your program. List and run your program.

Question 6. Attach the printout, marked P6, of your program for assigning and reading array values. Suggest a program for creating a 90-element array which would contain the values of the sines of the angles equal to the subscripts of the elements in degrees. How would you take advantage of this array to obtain the sine of a particular angle?

FOR-NEXT loops can be NESTED -- placed one inside another -- but the inner loop must be entirely within the outer loop. Such loops are often used with arrays of more than one dimension. For example, try:

```
NEW
10 DIM A(4,6)
20 FOR I=0 TO 4
30 FOR J=0 TO 6
40 A(I,J)=I*J+1
50 NEXT J
60 NEXT I
70 FOR I=0 TO 4
80 FOR J=0 TO 6
90 PRINT A(I,J);
100 NEXT J:PRINT
110 NEXT I
120 END
```

This program creates a (4+1) times (6+1) or 35 element array. These multidimensional arrays are useful when dealing with data for which more than one measurement is made at each time. The first subscript could be interpreted as the time, and the second subscript as the measurement type.

The ; in statement 90 changes the PRINT statement subtly. Previously, each item was separated from the previous item by a number of spaces so that the output appeared in tabular form. Also, after each PRINT statement, a line feed was performed on the printer. The ; causes no tabulation or line feed to occur. The result is a more compact form of output.

In line 100, the : is used to combine two separate statements, NEXT J and PRINT. This can be used to make the program more compact. The PRINT command with no quotation following is used to produce a line feed.

## 12-6   VARIABLE TYPES:   FLOATING-POINT, INTEGER, AND STRING

So far, all variables have been capable of storing real numbers in the range of about 10E-39 to 10E39.  In the AIM-65 version of BASIC, each such real or FLOATING POINT number requires seven bytes for storage:  two bytes for the variable name, one for the exponent, and four for the fraction.  It is difficult to store or interpret numbers in this format when doing digital input and output, so an alternative method of storage is provided.  This method, called INTEGER, stores the value as a 16-bit two's complement number. Integer variables have no fractional parts, and can represent numbers from -32767 to +32767.  BASIC converts integer variables to floating point prior to using them in calculations.

An integer variable name ends with a %.  Try the following:

```
NEW
10 INPUT A%,B%
20 C% = A% / B%
30 C = A% / B%
40 PRINT C%,C
50 GOTO 10
```

In the program above, GOTO 10 simply means that statement 10 should be executed after statement 40.  Note that C and C% are different variables:  C% is NOT the integer version of C.

Experimentally verify the effect of assigning various positive and negative real numbers to integer variables.  Also verify the range of values that integer variables can represent.  Integer variables are of interest primarily because sixteen bit values are often natural representations of data acquired from experiments.  In addition, the storage of integer arrays is particularly straightforward, as will be shown later.

Question 7.   Describe the effects of assigning various positive and negative real numbers to integer variables.  What is the effect of exceeding the magnitude limit for integer variables?  Attach the relevant printout, marked P7.

String variables are used to store strings of characters, such as might be used for titles or names.  A string variable name ends in $.  A number of functions are provided in BASIC for dealing with strings.  A list of these functions is provided below.

| | |
|---|---|
| LEN(string) | Returns the length (number of characters) in the string. |
| LEFT$(string,length) | Returns at the length leftmost characters of string. |
| RIGHT$(string,lenght) | Returns the length rightmost characters of string. |
| MID$(string,start,length) | Returns the length characters of string, beginning at the start character. |
| STR$(expression) | Converts a numeric expression to a string. |
| VAL(string) | Converts a string to the numeric value it represents. |

Notice that the name of any functions which produce a string ends in $. Only one "arithmetic" operation works with strings, namely +, which joins two strings to make a longer string.

Try the following:

```
A$ = "YOUR NAME"
PRINT A$
```

The first statement defines A$ and the second commands the string to be printed. Now try:

```
NEW
10 A$ = "YOUR NAME"
20 FOR N = 1 TO LEN(A$)
30 PRINT LEFT$(A$,N)
40 NEXT N
```

The expression LEN(A$) is a number equal to the number of characters in the string. LEFT$(A$,N) is the leftmost N characters in the string A$. Now run this program again, first replacing line 30 with RIGHT$(A$,N) and then with MID$(A$,N,3).

Question 8. Attach your printout for the three "YOUR NAME" programs (P8). Describe the portions of the string selected by RIGHT$(-,-) and MID$(-,-,-).

New strings can be defined from previous ones. For example, run:

```
NEW
10 A$="CAT's CRADLE"
20 B$=A$:  PRINT B$
30 L=LEN(A$) -1
40 FOR N=1 TO L
50 B$=RIGHT$(B$,L)+LEFT(B$,1)
60 PRINT B$
70 NEXT N
80 END
```

To demonstrate your grasp of these string concepts, write a program that will reverse the order of characters in a string of less than 20 characters. List and run your program.

Question 9. Explain your program and attach the printout (P9).

12-7  LOGICAL OPERATORS AND SUBROUTINES

AIM BASIC supports three logical operators which perform bit-wise manipulations on 16-bit integers. Floating-point numbers used with these operators are first converted to integer, if possible. The 16-bit result of these operations is converted to floating-point if the variable to which the result is assigned is floating-point.

The AND function is often used to select or mask certain bits of a word. It is useful to think of such masks as hexadecimal numbers, since the position of the bits can be more easily deduced. For example, in line 1020 of the program below, H% AND 15 yields the least significant four bits of H%. Note that 15 is "F" in hexadecimal, which is 1111 in binary.

AIM BASIC provides no facility for the input or display of hexadecimal numbers. Shown below is a program which prints the hexadecimal representation of the value stored in the integer variable H%. Enter and test the program. List the program.

```
NEW
10 A$="0123456789ABCDEF"
20 INPUT H%
1000 H$=""
1010 FOR H=1 TO 4
1020 H$=MID$(A$, (H% AND 15)+1,1)+H$
1030 H%=H%/16
1040 NEXT H
1050 PRINT H%,H$
1060 GOTO 20
```

Question 10.  Beside each statement, in the listing of the above program, write a brief description of what it does.  Attach the printout (P10).

A modification of the program will allow reuse of the code to print the hexadecimal form of an integer variable.  Without typing NEW, enter the following statements:

```
20 INPUT A%,B%
30 H%=A%:GOSUB 1000
40 H%=B%:GOSUB 1000
50 H%=A% AND B%:GOSUB 1000
60 H%=A% OR B%:GOSUB 1000
70 H%=NOT A%:GOSUB 1000
80 GOTO 20
1060 RETURN
```

List the modified program.  The statements 1000-1060 are now in the form of a SUBROUTINE.  Each time the subroutine is CALLED, by the GOSUB 1000 statement, the BASIC interpreter notes where the call occurs, jumps to statement 1000, and executes statements 1000-1060.  Statement 1060 causes the interpreter to return to the statement after the statement that called the subroutine.  The : simply allows two statements to be placed on one line, usually for ease in reading.

Run the program, using a variety of different values for A% and B%.

Question 11.  Submit the listing and results (P11).  Describe an application where subroutines could be used to advantage.  How could the EXCLUSIVE-OR function be accomplished in BASIC?

A.  IF ... THEN ...

It is usually necessary to alter the order in which program statements are executed based upon characteristics of the data, or other variables. BASIC provides a conditional branch in the form of the IF ... THEN ... statement.  Enter the following example:

```
NEW
10 INPUT "A";A
20 INPUT "B";B
30 IF A<B THEN PRINT "A<B"
40 IF A>B THEN PRINT "A>B"
50 IF A=B THEN PRINT "A=B"
60 IF A<=B THEN PRINT "A<=B"
70 IF A>=B THEN PRINT "A>=B"
80 IF A<>B THEN PRINT "A<>B"
90 GOTO 10
```

The RELATIONAL operators =, <, >, <>, >=, and <=, evaluate to a value of 0 if the indicated relation is false and to -1 if the indicated relation is true.  The value -1 was chosen since its 16-bit representation is 1111111111111111.  In fact, the THEN clause will be executed if the expression in the IF clause evaluates to any non-zero number.  The THEN clause may contain multiple statements separated by colons.  In that case, all the statements will be executed if the IF clause expression is true.  Otherwise, the next NUMBERED statement will be executed.

The following program fills a 20 element array with random numbers between 0 and 1000 and picks out the largest and smallest numbers in the array.  The command RND(1) generates a new random 9-digit fraction. Multiplying this fraction by 1000 and taking the integer portion (by the INT( ) function) gives integers between 0 and 999.  List and run this program.

```
NEW
10 DIM A(20)
20 FOR N=1 TO 20
30 A(N)=INT(1000*RND(1))
40 PRINT A(N)
50 NEXT N
60 B=A(1):C=A(1)
70 FOR I=1 TO 20
80 IF A(I)>B THEN B=A(I)
90 IF A(I)<C THEN C=A(I)
100 NEXT I
110 PRINT "THE LARGEST # IS",B
120 PRINT "THE SMALLEST # IS",C
```

Question 12.  Attach your printout (P12) and explain the section of the program that picks up the largest and smallest numbers.  What is the probability you will find a 999 in any given run?  How would you change the program to produce 4-digit numbers and select from among 100 random numbers?

Write one of the three following programs:

A. A program to generate 25 random numbers between 0 and 9999 and arrange them in increasing order.
B. A program to arrange 5 first names in alphabetical order (use comparisons of strings).
C. A program to rearrange any name given in the form FIRST LAST to the form LAST, FIRST.

The expressions in the IF clause can become extremely complex. An example is:

```
10 IF A>B AND C=<D OR G 2-3<5 THEN PRINT H
```

The unambigous interpretation of this IF clause is performed by the application of operator PRECEDENCE. The rules of precedence for AIM-65 BASIC are as follows:

1. Expressions in parenthesis
2. Exponentiation
3. Negation
4. Multiplication and division, as encountered from left to right
5. Addition and Subtraction, as encountered from left to right
6. Relational operators, as encountered from left to right
7. NOT
8. AND
9. OR

## 12-8 EXECUTION TIMES AND SPEED HINTS

In this section the speed of the various BASIC arithmetic operations will be measured, with a view towards increasing the speed of execution of BASIC programs. Enter the following program.

```
NEW
10 A=1234567.
20 B=.0123456
30 C=1
40 FOR I=1 TO 1000
50 D=C
60 NEXT I
70 END
```

Run the program, measuring the time between the pressing of the RETURN key and the end of the program. Record the time on the output. Then measure and record the time required for the program to execute with each of the following statements as line 50:

```
50 E=A*B
50 E=1234567*.0123456
50 E=A*A
50 E=A^2
50 E=A+A
50 E=2*A
50 E=2.00000000*A
50 E=A+B
50 E=B+B
50 E=A/B
```

Note that the times are consistent with the fact that the BASIC interpreter must translate the each line each time it is executed. Also note that it is faster to find the value variable A and B than it is to translate the numbers from decimal to BASIC's internal form.

## 12-9  BASIC – MACHINE LANGUAGE INTERFACE

As has become apparent in the speed tests in Part I, the rate at which BASIC could possibly acquire data is insufficient for many processes. One way of resolving this difficulty is to use machine language subroutines called from a BASIC program. In order to do this, an understanding must be acquired of the way in which BASIC stores variables, paticularly integer arrays. This can be readily accomplished using the MONITOR program which comes with the AIM-65.

### A.  BASIC Array Storage and Examination of Memory

The actual location of the array A% depends upon the size of the BASIC program in memory, the number of other variables defined, and the order in which the other variables are defined. In order to keep track of where everything is stored, BASIC maintains POINTERS which contain the addresses of various important areas of memory. For now, the most important is the pointer to the start of array storage, which is located at $77 and $78 ($xx means xx is hexadecimal). The next section of this experiment will create a BASIC array, and examine how it is stored in memory.

To begin this section, it is important that the machine be re-initialized. Press the RESET button. Enter BASIC and type the following:

```
NEW
DIM A%(10,2)
```

There are two methods by which one can examine and alter the contents of memory in the AIM-65. The first method works directly from BASIC, using PEEK and POKE. PEEK is a function which returns the value of the byte whose address is the expression in the parenthesis. For example, to examine the contents of the array storage pointer, try the following:

```
PRINT PEEK(120)
PRINT PEEK(119)
PRINT PEEK(120)*256+PEEK(119)
```

The pointer to the start of array storage is a 16-bit word which is stored in two 8-bit bytes. The order of storage is called BYTE REVERSED, since the least significant byte has the lower address. In this case, location 120 has the most significant byte. In order to create the entire 16-bit word, it is necessary to shift the most signifcant byte 8-bits to the left, which is equivalent to multiplying by 256. Thus the last PRINT above prints the address of the first storage location in the array storage.

The array storage area is always just above the non-array storage area. This means that defining any other variables will move the array storage area. For example:

```
A=3
PRINT PEEK(120)*256+PEEK)119)
```

This means that care must be exercised to make sure the current array storage pointer is used. The array storage array can be examined by running the following program:

```
10 DIM A%(3,2)
20 I=0
30 J=0
40 S=PEEK(120)*256+PEEK(119)
50 FOR I=0 TO 3
60 FOR J=0 TO 2
70 A%(I,J)=I*4+J
80 PRINT I;J;A%(I,J)
85 NEXT J:NEXT I
90 PRINT "HEADER"
100 FOR I=S TO S+8
110 PRINT PEEK(I)
120 NEXT I
125 PRINT "DUMP"
130 FOR I=S+9 TO S+24+8 STEP 2
140 PRINT PEEK(I);PEEK(I+1)
150 NEXT I
160 END
```

This program defines all variables it uses in statements 10-30 so that the storage area will not change later in the program. The loops at 40-80 fill the A% array with unique numbers. The limits on the 110-130 loop are the starting address of array storage, S, and the S plus the number of elements in the array (3+1)*(2+1) times the number of bytes per element -- two -- plus 8, which is the number of bytes in the ARRAY HEADER minus one. The array header is used by BASIC to store the name of the array, its length, and the number and size of the subscripts. The values of the A% array are stored, 2 bytes per element, in the 10th through 33rd bytes of the array.

Question 13. Match the bytes printed in the 110-130 loop to the array elements printed in the 40-80 loop. Give the order in which the array elements are stored in memory [A(0,0), A(1,0), etc.].

While it is relatively easy to see the array elements in the storage area, it is useful to examine the header area in hexadecimal. This can be accomplished using the MONITOR program in the AIM-65.

The next instructions assume that you have entered and run the program above immediately before you start the next steps. Enter the monitor by pressing ESC. The monitor prompt character, <, shows at the left of the display. To examine memory, the M command is used. Press M. The display shows:

<M>=

Type 0077 and RETURN. The display and printer will show the contents of location 0077 and the next three locations (0078, 0079, 007A).

<M>=0077 1D 03 3E 03

The pointer to the start of array storage is stored in byte-reversed order; the first entry, 1D, is the least significant byte, and next entry, 03, is the most significant byte. Thus storage begins at $031D in this example. The exact location at which storage begins depends upon exactly how you entered the program: where you put spaces, etc. This is because storage begins after the end of the program. Therefore your pointer in locations $77 and $78 may not be 031D. Use whatever value you have to examine the storage area. Again press M and respond with your address and RETURN. To view the next four locations, press the space bar. Do this 10 times. A listing similar to the one below should result, although your addresses may differ.

```
<M> = 0077 23 03 44 03
<M> = 0323 C1 80 21 00
<  >   0327 02 00 03 00
<  >   032B 04 00 00 00
<  >   032F 04 00 08 00
<  >   0333 0C 00 01 00
<  >   0337 05 00 09 00
<  >   033B 0D 00 02 00
<  >   033F 06 00 0A 00
<  >   0343 0E AA AA AA
<  >   0347 AA AA AA AA
<  >   034B AA AA AA AA
```

The first two bytes of the header encode the name and type of the array. The least significant seven bits of these bytes contain the ASCII code for the two character array name. Both of the most significant bits are set to 1, indicating that this is an integer array. (The other three combinations represent floating point, string, or functions definition.) The next two bytes are the length of the storage area for this array, in a byte-reversed 16-bit form. The next byte is the number of subscripts used in the dimension statement. The next pairs of bytes are the subscript values, which are one larger than the corresponding dimension because of the zeroth element. These numbers are NOT byte-reversed. The remaining pairs of bytes are the actual array values, also not byte-reversed.

Question 14. Rewrite the BASIC program to list the elements in the array in the same order that they are stored in memory.

B. Calling Machine Language Routines -- the USR Function

In this section a method of calling machine language routines from BASIC is developed. Procedures similar to this are used in interfacing experiments in later Units.

The USR function causes the computer to begin execution of the machine language program that begins at the location pointed to by the contents of bytes 4 and 5. The machine language program must be set up in an area of memory not used by BASIC. To reserve space for these routines, the memory size question that appears when BASIC starts must be answered with a number instead of a return. This is done in the example below. An example USR routine is implemented in an exercise below. But first, it is instructive to locate some of the significant pointer addresses in the execution of a USR routine.

When the USR routine is called, BASIC places the value of the expression in parentheses in a special set of memory locations called the floating-point accumulator. A routine can be used to convert this value to a 16-bit integer. The address of this routine is stored in locations $B006 and $B007 in byte-reversed order. Use the M command from the monitor to examine the memory locations $B006 and $B007. Record the address found there in the space below.

FP to INTEGER ROUTINE ADDRESS: _____

The integer value thus produced is stored in locations $AC (most significant byte, or MSB) and $AD (least significant byte, or LSB).

The 6502 microprocessor used in the AIM-65 has three 8-bit registers. The A register, or accumulator, is the primary computational register. The X and Y registers are used primarily for temporary storage and for a variety of indexed addressing operations. When a machine language subroutine ends and returns to BASIC, the 16-bit integer stored in the A register (MSB) and in the Y register (LSB) is converted to floating-point and becomes the value of the USR function. The address of this routine is stored in locations $B008 and $B009. Use the M command to obtain this address, and record it below.

INTEGER to FP RETURN ROUTINE ADDRESS: _____

The following examples illustrate and clarify these points in USR function operation. The first USR function causes the hexadecimal equivalent of the 16-bit integer value of the expression in the parenthesis to be printed. It does this by using the FP to INTEGER routine, a monitor routine called NUMA, and the RETURN routine. The NUMA routine prints the contents of the A register as two hexadecimal digits.

Try to follow the sequence shown exactly.

1. Reset the computer by pressing the RESET button.

2. Enter BASIC by pressing 5. Answer 3967 to the MEMORY SIZE? question; press RETURN when asked WIDTH?.

On a 4K AIM-65, it is easiest to reserve about 128 locations by entering 3967 in answer to the memory size question. Memory locations $0F80 to $0FFF (3967 + 128 = 4095) will then not be used by BASIC.

3. Press ESC. The MONITOR prompt, >, now appears.

4. The machine language program will now be entered into the memory at $0F80. This is done using the monitor I command. This MNEMONIC MODE or I-mode entry allows the entry of machine language instructions in mnemonic form, rather than the absolute numeric form sometimes required. Enter the underlined characters in the listing below. Where underlined and non-underlined characters alternate on a single line, the computer has responded to a single character input with more than one character. Note that the information after and including the ; on each line are only present for clarification of the program, and cannot and should not be entered.

```
<I>
0000        *=0F80          ;SET STARTING ADDRESS
0F80  20  JSR BEFE[1]       ;USE FP to INTEGER ADDRESS: SET UP INTEGER VALUE
0F83  A5  LDA AC            ;PUT MSB IN REGISTER A
0F85  20  JSR EA46          ;CALL NUMA ROUTINE
0F88  A5  LDA AD            ;PUT LSB IN REGISTER A
0F8A  20  JSR EA46          ;CALL NUMA ROUTINE
0F8D  4C  JMP COD1[1]       ;RETURN TO BASIC
0F90  (press ESC)
```

Footnote 1. If the addresses you found at locations B006-B007 and B008-B009 were not BEFE and COD1, use the addresses you obtained in these instructions.

Press ESC to end the I-mode input.

5. Re-enter BASIC by pressing 6 (NOT 5). Pressing 6 causes BASIC to restart without resetting the memory size or erasing any program currently stored in the memory.

6. Enter and run the following BASIC program. Experiment with a variety of different input values.

```
10 POKE 4,128:POKE 5,15
20 INPUT "DECIMAL";D
30 A=USR(D)
40 PRINT
50 GOTO 20
```

Statement 10 sets locations 4 and 5 to point to $0F80 (128 = $80; 15 = $0F), which is the address of the USR routine. Statement 40 provides a carriage return and line feed, which NUMA, called through USR, does not. Note that USR is a function, and therefore must be used as part of an expression; a statement such as

USR(D)

will produce an error message.

A second, slightly more complex example is shown below. Once again, the machine language program is loaded at $0F80. The procedure is the same except for the program loaded in I-mode and the BASIC calling program. This program prints the hexadecimal equivalent of the values stored in positions 1 through N of the first array in the array storage area, which is assumed to be a single dimension integer array. The number of elements to be printed is requested in an input statement. The machine language program to be entered is

```
<I>
NNNN *=0F80
0F80 20 JSR BEFE        ;Get integer value of USR expression
0F83 18 CLC             ;Clear link bit
0F84 A5 LDA 77          ;Load LSB of array pointer
0F86 69 ADC #07         ;Add 7 to array pointer
0F88 85 STA FE          ;Store in location $FE
0F8A A9 LDA #00         ;Load Register A with 00
0F8C 65 ADC 78          ;Add MSB of array pointer, and carry if any
0F8E 85 STA FF          ;Store in location $FF
0F90 A0 LDY #00         ;Load index register Y with 00

0F92 B1 LDA (FE),Y      ;Load register A with the contents of the location
                        ;whose address is the sum of the contents of
                        ;register Y and the contents of locations $FE and
                        ;$FF.  This is the MSB of the current array element.
0F94 20 JSR EA46        ;Call NUMA routine
0F97 C8 INY             ;Increment register Y
0F98 B1 LDA (FE),Y      ;Repeat for LSB of array
0F9A 20 JSR EA46
0F9D C8 INY
0F9E 20 JSR E9F0        ;Do a RETURN and line-feed
0FA1 C6 DEC AD          ;Decrement number of elements to print
0FA3 D0 BNE 0F92        ;If not zero, repeat
0FA5 4C JMP C0D1        ;Return to BASIC
0FA8 (press ESC)
```

The BASIC program is:  (Remember to reenter BASIC with a 6)

```
            NEW
            10 DIM A%(25)
            20 FOR I=1 TO 25          Loads the 1st 25 positions
            30 A%(I)=I*I             in the array storage with
            40 NEXT I                the square of the number of
            50 POKE 4,128:POKE 5,15  the element.
```

```
60 INPUT "HOW MANY";N
70 I=USR(N)
80 PRINT "QUITS"
90 END
```

<u>Printout 14.</u>  Run the program to verify that it works properly, and submit the printer output as documentation.

<u>Question 15.</u>  Alter the program so that requests for zero elements or more than 25 elements are not accepted and an informative message is issued. Attach your printout (P15).


12-10  INTERRUPT ROUTINES

It is often useful to allow an external device to demand service from the computer even if the computer is otherwise occupied. The example explored in this experiment is the use of the computer as a clock. A timer is set up to emit a pulse every 50 milliseconds. The computer keeps track of the time by decrementing a memory location which was initially set to 20. When this location goes to zero, it is reset to 20, and a second location is incremented. This second register counts seconds. A third register is incremented each time the second register overflows, which occurs every 256 seconds. The trick is to avoid the undesirable waste of computer time which would occur if the computer had to watch continuously for the next pulse. This is an example where an interrupt driven system is useful. Each time the pulse occurs, the computer is forced to stop what it was doing, "service" the interrupt, and then pick up where it left off.

The program you are about to enter has four parts. The simplest is the BASIC program, which simply sets up the USR address, calls USR(0) to reset the clock, and then repeatedly calls USR(1) to obtain the number of seconds since the last USR(0) call.

Before entering the BASIC program, reset the computer, and enter BASIC with the 5 key. As before, reserve 128 locations by entering 3967 to MEMORY SIZE?.

```
NEW
10 POKE 4,160:POKE 5,15
20 I=USR(0)
30 PRINT USR(1)
40 GOTO 30
50 END
```

There are three sections to the machine language code for this experiment. The first, at location $F80, is the interrupt service routine. The interrupt service routine is written in machine language because it must execute quickly and because there is no facility in this version of BASIC to handle interrupts. Enter this routine using the mnemonic entry or I-mode:

```
<I>
0000      *=F80
0F80 48 PHA
0F81 CE DEC 0FFF
0F84 D0 BNE 0F93
```

```
0F86 A9 LDA   14
0F88 8D STA  0FFF
0F8B EE INC  0FFE
0F8E D0 BNE  0F93
0F90 EE INC  0FFD
0F93 AD LDA  A004
0F96 68 PLA
0F97 40 RTI
0F98 (press ESC)
```

This routine is very simple. First, the contents of register A are stored on the stack, so that the A register can be used without fouling up the program that was interrupted. The contents of location $FFF are decremented. If the contents are still non-zero, the routine is exited. If the contents are zero, they are reset to 20 ($14 in hexadecimal), and the contents of the least significant byte of the seconds counter at address $FFE are incremented. If this does not result in an overflow, the routine is exited. If an overflow occurs, the MSB of the seconds counter, at location $FFD, is incremented. The exit consists of reseting the interrupt flag on the pulse generator, restoring the value of the A register, and returning from the interrupt (RTI).

The third and fourth sections are combined in one USR machine language routine.

```
              <I>
0F98     *=FA0
0FA0 20 JSR  BEFE
0FA3 A5 LDA  AC
0FA5 05 ORA  AD
0FA7 D0 BNE  0FD8
0FA9 A9 LDA  #80
0FAB 8D STA  A400
0FAE A9 LDA  #0F
0FB0 8D STA  A401
0FB3 A9 LDA  #00
0FB5 8D STA  0FFD
0FB8 8D STA  0FFE
0FBB A9 LDA  #14
0FBD 8D STA  0FFF
0FC0 58 CLI
0FC1 A9 LDA  #C0
0FC3 8D STA  A00E
0FC6 A9 LDA  #40
0FC8 8D STA  A00B
0FCB A9 LDA  #64
0FCD 8D STA  A004
0FD0 A9 LDA  #C3
0FD2 8D STA  A005
0FD5 4C JMP  COD1
0FD8 AD LDA  0FFD
0FDB AD LDY  0FFE
0FDE D0 BNE  0FEB
0FE0 AD LDA  0FFD
0FE3 4C JMP  COD1
0FE6 (press ESC)
```

In this case, the value of the expression in parenthesis determines what the USR routine does. The subroutine at BEFE converts the expression to a 16-bit integer stored in $AC and $AD. The contents of $AC are loaded into the A register, and logically ORed with the contents of $AD. If the result is non-zero, the routine reads the current state of the second counter, as described below. If the result is zero, the clock is reset. The following operations are performed:

1. The address of the interrupt service routine, $F80, is loaded into memory locations $A400 and $A401. This is where the computer looks for an address when the interrupt occurs. The process used to get to that address is too complex to explain in the space available here.
2. The seconds counter at $FFD (MSB) and $FFE (LSB) is cleared. The 50 millisecond pulse counter at $FFF is set to 20 ($14 hexadecimal).
3. The computer is enabled to accept interrupts (CLI instruction).
4. The next eight instructions set up the pulse generator which is part of 6522 Versatile Interface Adapter, which will be studied in Unit 13.
5. The routine exists to BASIC. No particular value is set.

Reading the counter is accomplished by the code at $FEO and following. First the MSB (most significant Byte) of the seconds counter is read and stored in the A register. The LSB is then read, and stored in the Y register. If the LSB is zero, the MSB is read again. The routine at COD1, in addition to returning to BASIC, also converts the 16-bit integer value in the A register (MSB) and Y register (LSB) to floating-point. This becomes the value of the USR function.

Question 16. Why is the MSB re-read if the LSB is zero? (Hint: What could have happened between the reading of the MSB and the reading of the LSB?)

Reenter BASIC after entering the machine language code by using 6. Turn the printer on the AIM-65 off by pressing PRINT while holding the CTRL key down. This will reduce the paper waste. Run the BASIC program. Once started, the interrupt routine continues in operation even if the BASIC program has been stopped. Note the time on the display. Press F1. Then type:

PRINT USR(1)

Observe that the clock is still running. You can now enter other programs, alter the program in memory, etc., and the clock keeps ticking away.

Question 17. Write a BASIC program that prints the time in hours, minutes and seconds. You will need to ask what time it is, reset the clock using USR(0), and then compute the current time from the USR(1) value and the starting time. Attach your printout (P17).

Question 18.  Show how to convert this program to a subroutine that could have been used to measure the time taken by the 1000 cycle loop in the program used to determine excution speed of BASIC statements.

The only way to kill the clock is to reset the computer, which you should do before continuing with other experiments.

# UNIT 13.  INTRODUCTION TO INTERFACING

This unit introduces the most important concepts of microcomputer interfacing.  The 6522 Versatile Interface Adapter (VIA) is used in most of its many modes.  The VIA contains two 16-bit counter/timers, two 8-bit input/output ports, a shift register, and 4 additional input/output lines which can be used in conjunction with the other elements.  A single chip with these capabilities is necessarily complex.  In view of this, the experiments usually exercise only one feature at a time.

The goal of the unit is not solely to understand the operation of the 6522, but rather to introduce the basic concepts of microprocessor interfacing.  Basic parallel input and output will be demonstrated.  The counters on the 6522 will be used to build a simple but versatile frequency/period meter.  A digital-to-analog converter will be connected to the microprocessor, and used to demonstrate various types of servo-controlled analog-to-digital converters.  An integrated successive approximation analog-to-digital converter will be exercised.  Finally, a complete data-logging system will constructed.

| Equipment | Parts |
|-----------|-------|

1.  Basic station
2.  AIM-65 microcomputer
3.  Interface job board
4.  BSI job board
5.  Cassette recorder

## 13-1  INTRODUCTION TO THE VERSATILE INTERFACE ADAPTER

Objectives:  To introduce the AIM-65 interface job board and perform simple parallel input and output operations.

### A.  The 6522 Interface Job Board

The AIM-65 microprocessor system has a 6522 Versatile Interface Adapter, or VIA, designated for operator use.  The interface pins of this device are connected to plug J1 on the back of the AIM-65.  The 6522 VIA interface job board and the cable used to connect it to the 6522 VIA are shown in figure 13-1.  For convenience, the interface job board should be placed in the right-most position in the mainframe, and the AIM-65 should be to the right of the mainframe.

240

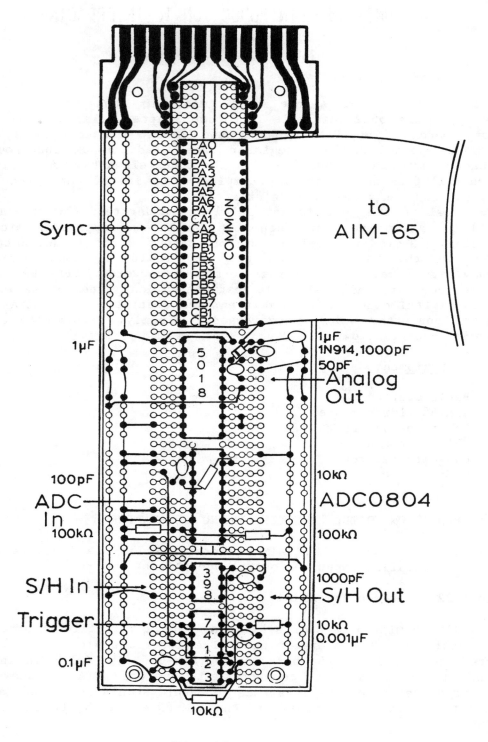

Fig. 13-1

The outputs of the 6522 which come to the interface board consist of two sets of eight lines and two sets of two lines. The pin designations and their location on the interface cable header installed on the interface job board are also shown in figure 13-1. Note that ALL the pins on the cable side of the connector (pins numbered 21-40) are connected to the AIM-65 common. A connection between one of these pins and the mainframe common must be made, as shown. The pin designations will be further clarified as the experiments are performed; for now note that PA0-PA7 are one eight bit group, and PB0-PB7 are a second eight bit group. The control lines CA1 and CA2 are associated with the A port (PA0-PA7), while CB1 and CB2 are control lines for the B port.

NOTE: Throughout this experiment, all numbers are decimal, except when preceded by a $ for hexadecimal. thus 40 is decimal, and $40 is $40_{16}=64_{10}$.

B. General Organization of the 6522 VIA

The 6522 VIA is controlled by reading from and writing into the sixteen registers it contains. On the AIM-65 system, the sixteen registers are addressed as sixteen consecutive memory locations beginning at $A000 or 40960. The list of locations and the functions they control is shown in the AIM-65 Summary Card. Basically, there are two data registers (ORA and ORB), a shift register (SR), two timers (T1 and T2), two data direction registers (DDRA and DDRB), and four control registers: the auxiliary control register (ACR), the peripheral control register (PCR), the interrupt flag register (IFR), and the interrupt enable register (IER). Each of these registers will be introduced when needed.

C. Basic Parallel Input/Output

The simplest modes of operation of the 6522 VIA involve its use as a parallel input/output device. Connect LEDs A - H on the BSI job board to VIA pins PB0-PB7 (interface header pins 11-18). Connect switches A-H on the BSI job board to VIA pins PA0-PA7 (interface header pins 1-8). RESET the computer, enter BASIC, and enter the following program:

```
NEW
10 A=40960
20 POKE A+3,0
30 POKE A+2,255
40 S=PEEK(A+1)
50 L= NOT S AND 255
60 PRINT S;L
70 POKE A,L
80 GOTO 40
```

As mentioned above, the VIA addresses start at 40960 or $A000. Assigning this to variable A makes for less typing. The register at A+3 is the port A DATA DIRECTION REGISTER (DDRA). The contents of each bit of this register determine whether the corresponding bit of the A port acts as a input or an output. If the bit in the DDRA is 1, the corresponding bit in the port is an output. A 0 in the DDRA means that bit is an input. It is possible to split one port between inputs and outputs. The register at A+2 is the data direction register for port B (DDRA). It functions in an exactly analogous manner. The program sets all of port A to input and all of port B to output. At 40, the BSI switch settings are read through port A and the value assigned to S. The logical complement is taken, both values printed, and the complement is sent to the B port pins. Turn off the printer (press CTRL and PRINT together) to conserve paper. Run the program, and verify that it works.

Question 1. Why is the AND 255 required in statement 50 in the program above?


13-2  SIGNALLING AND HANDSHAKING

Objectives:  To develop methods of synchronizing the computer with external events, and to demonstrate the technique of handshaking.

The program above requires the computer to spend all of its time PEEKing and POKEing. It performs these operations even when nothing has changed, or during changes that are not complete. Methods will now be demonstrated which allow the computer to synchronize itself with the outside world.

The control lines CA1, CA2, CB1, and CB2 are used both by the outside world to signal that something needs the computer's attention and by the computer to start events or announce new output. The CA1 and CA2 pair behave with respect to A port almost identically to the way CB1 and CB2 behave with respect to B port. The following discussion will first focus on the A port.

CA1 is always an edge-triggered input. PCR0, the least significant in the PERIPHERAL CONTROL REGISTER (PCR – address $A00C or 40972) determines whether the CA1 is rising-edge triggered (PCR0 = 1) or falling-edge triggered (PCR = 0). When an active transition occurs, the event is recorded in the INTERRUPT FLAG REGISTER (IFR – address $A00B or 40971). The IFR consists of eight one-bit flags, seven of which indicate activity of a specific type on their associated device. IRF1 is set to 1 when an activating edge occurs on CA1.

Without unwiring anything from above, connect the BSI job board MA to CA1 (interface header pin 9). RESET the computer, enter BASIC, and enter the following program:

```
NEW
10 A=40960
20 POKE A+12,1
30 POKE A+13,255
40 PRINT "PUSH WHEN READY"
50 IF (PEEK(A+13) AND 2) = 0 THEN 50
60 PRINT "THANK YOU"
```

Run the program. Indicate on which edge the flag is raised. Change statement 20 to POKE A+12,0. Run the program again and observe the change in the edge sensitivity.

Observations:

Statement 50 can be replaced by a WAIT statement. This responds much more quickly than the equivalent BASIC statement. The WAIT statement has three arguments. The first is the address of the flag byte which is to be read and checked. The second is the byte value that is ANDed with the flag byte. The final parameter is exclusive ORed with the flag byte before the AND operation. If the third argument is not present, it is assumed to be zero.

Thus the statement:

```
50 WAIT A+13,2
```

does the same as the statement at 50 in the program you just ran.

Replace statement 50 with the WAIT statement above. Verify that the program performs the identical function as the previous program. Note that to stop this program, one must press RESET and reenter BASIC by pressing "6". This is because the computer ignores the keyboard during the execution of a WAIT statement.

The CA1 flag can be cleared either by reading or writing port A ($A001 or 40961) or by directly writing a 1 (yes, really a 1 to clear) into the IFR1 bit position. Unfortunately, the state of the IFR flag cannot be detected outside the computer, with the result that it is not possible to tell whether the computer has actually reacted to the request. A more sophisticated method of signalling, known as HANDSHAKING, can be employed. In this case, CA2 is set to 1 by an active transition on CA1, and cleared by a read or write of the A port To enable this mode, the appropriate bits in the PCR must be set: PCR1=PCR2=0; PCR3=1. In addition, PCR0 will be set to activate CA1 on a positive edge. Connect CA2 (header pin 10) to an LED on the BSI job board. Change 20 to:

```
20 POKE A+12,8+1
```

Add the following statements, which simulate something the computer might need to do before actually taking the data, and which provide time to observe the state of CA2 between the edge an CA1 and the read operation:

```
70 FOR I=1 TO 1000:NEXT I
80 PRINT PEEK(A+1)
90 GOTO 40
```

List the modified program. Run the program, noting on the tape the state of CA2 between the time PB1 is pressed and the time statement 80 is executed.

Printout 1. Attach the above listing and execution tape marked P1.

Port A will operate in the handshaking mode for both read and write (PEEK and POKE) operations. The analogous mode in port B operates only for write operations. As shown in the Summary Card, PCR5-7 control CB2; PCR4 controls CB1.

One other problem exists. It is possible that the data present on the input lines of port A will change between the time the computer is signalled and the time the data are read. A useful method of avoiding this problem is to latch the data into the port on the active edge of CA1. This operation is controlled by a bit ACR0 in the AUXILIARY CONTROL REGISTER (ACR - address $A00B or 40971).

Run the above program again. Change the position of one or more BSI switches quickly after pressing PB1. Note that the modified value is the one printed.

Now insert the following modification, which sets the latching mode on CA1.

35 POKE A+11,1

Run the program again to demonstrate that it is now impossible to fool the computer.

Disconnect all switches and LED's from the interface header.

There are a number of other modes in which CA2 (and analogously, CB2) can operate. If PCR3=0, CA2 is an input. Under this condition, PCR2=0 sets positive edge triggering, and PCR2=1 sets negative edge triggering. PCR1=0 allows the IFR0 flag bit to be cleared either by reading or writing port A or by writing IFR0=1. PCR1=1 allows the IFR0 flag bit to be cleared only by directly writing to the IFR.

If PCR3=1, CA2 is an output. The four states of PCR2 and PCR1 are:

1. PCR2=PCR1=0: Handshake mode, discussed above.
2. PCR2=0,PCR1=1: Pulse mode: CA2 goes low for one pulse each time port A is read or written.
3. PCR2=1,PCR1=0: Set CA2 low.
4. PCR2=PCR1=1: Set CA2 high.

The CB2 control bits behave identically. See the Summary Card.

Question 2. Suggest a use for each of the output modes of CA2.

13-3 THE 6522 TIMER/COUNTERS

Objectives: To become familiar with the 6522 timers T1 and T2, and to use them to build a TTL frequency generator and a frequency meter.

A. Introduction to T1

The 6522 has two timers, T1 and T2. The first part of this experiment will deal with T1 exclusively. T1 consists of a sixteen bit counter and a sixteen bit latch. Since the 6502 microprocessor in the AIM-65 is an eight bit machine, the sixteen bit registers are divided into two bytes each. The counter bytes are T1C-L and T1C-H for T1 Counter, Low order byte and T1 Counter, High order byte, respectively. Similarly, the latch bytes are called T1L-L and T1L-H.

From the AIM-65 Summary Card section on R6522 Memory Assignments it is clear that four addresses, from $A004 through $A007, affect T1. The effect of writing a byte into the various addresses is not immediately obvious. None of the commands write directly into the counters; all write into the latches. Addresses $A004 and $A006 behave identically when writing; both load the low byte of the latch. Writing into $A007 loads the high byte of the latch, and clears the T1 flag in the interrupt flag register (IFR). Writing into $A005 also loads the high byte of the latch, and clears the T1 bit in the IFR. But writing into $A005 also loads the counter bytes from the corresponding latch bytes, and starts the counter counting down. What happens next depends upon the state of ACR6 and ACR7, which control the T1 mode selection.

In the T1 one-shot mode (ACR7=ACR6=0) the T1 flag in the IFR is set N+2 clock cycles after the counter is loaded with the number N. If ACR7=1 and ACR6=0, the mode is still T1 one-shot, but now the PB7 output pin of the 6522 goes low when the counter is loaded and stays low until the T1 flag is raised. In both these modes, the counter continues to count through zero, but no further changes in the T1 flag or the output pin, if active, occur until the counters are reloaded. The other two modes are free-running modes, in which the counter is reloaded from the latches each time the counter reaches zero. If ACR7=0 and ACR6=1, the T1 flag is set each time the counters are reloaded. No output occurs. If ACR7=1 and ACR6=1, the PB7 output goes low when the counters are first loaded, and then changes state each subsequent time the counters are loaded. As above, the T1 interrupt flag in the IFR is set each time the counters are reloaded. The clock signal used to decrement the counters is the approximately 1 MHz system clock of the AIM-65.

Both the counter and the latch can be read by PEEKing the appropriate addresses. A PEEK at address $A004 returns the low byte of the T1 counter. This operation also clears the T1 flag in the IFR. Address $A005 returns the high order latch byte. PEEKing in addresses $A006 and $A007 returns the low and high order byte, respectively, of the counter. Note that the value returned is the instantaneous value: if the counter is still counting, the value may change during the time it takes to read both bytes.

This can all be demonstrated more readily than it can be explained. Enter the following program:

```
NEW
10 A=40960
20 POKE A+11,128+64
30 INPUT "COUNT ";C
40 POKE A+4,INT(C-(INT(C/256)*256)
50 POKE A+5,INT(C/256)
60 INPUT "FREQ";F
70 INPUT "PERIOD";P
80 GOTO 30
```

Statement 20 sets ACR7=ACR6=1, which means T1 is in the free-running output mode. Statement 30 acquires the counter value, which is loaded into the low byte in statement 40, and into the high byte in statement 50. Loading at A+5 ($A005) also starts the counting. Connect the PB7 output of the VIA (header pin 18) to the scope and to the CTFM.

246

Record on the printer tape the frequency and period observed for count values of 1, 2, 5, 10, 100, 1000, 10000, 65535, and 0.  Try 65536 or any negative value, and observe the result.

Result _____

<u>Question 3.</u>  How many "extra" counts does the 6522 add to the C value?  Attach your printout (P3).

It is worth noting that asymmetrical "square" waves can be generated by reloading the latches before the counter has counted down to zero.  This is difficult in BASIC, however, since BASIC takes so long to execute the reload functions.

Write a program that accepts the frequency desired, calculates the appropriate count value, starts the counters, and prints the actual frequency output.  If the desired frequency is out of the range available, issue a warning message.  Demonstrate that the program works.

<u>Printout 4.</u>  Attach the printout for the above program.

B.  Single Shot Operation

The one-shot output mode of T1 will be demonstrated with the following program:

```
NEW
10 A=40960
20 POKE A+11,128
30 INPUT "COUNT ";C
40 CL=INT(C-INT(C/256)*256)
50 CH=INT(C/256)
60 POKE A+4,CL
70 POKE A+5,CH
80 GOTO 60
```

In order to measure the duration of the low pulse on PB7, the time A-B mode of the CTFM must be used.  Connect PB7 to A input of the CTFM and to the input of an inverter.  The inverter output is connected to the B input of the CTFM.  Also connect PB7 to the scope.

Run the program.  On the tape, record the pulse width for counts of 1, 2, 5, 10, 100, 1000, and 10000.  Also record, from the scope, the time between the start of successive pulses.

<u>Printout 5.</u>  Attach the printout.

Notice that for the larger count values, the pulses never end.  This is because reloading restarts the counters.  Thus a WAIT statement is needed.  The T1 flag is bit 6 of the IFR, so a mask value of 64 is required.  Therefore, insert:

```
75 WAIT A+13,64
```

Repeat the above exercise, and note any differences.  Determine the minimum and maximum pulse widths available.  Record your results below.

minimum _____

maximum _____

Note that the PB7 pin of the VIA will reflect the T1 output even if port B is otherwise assigned as an input of if port B is used as a parallel output.

## C.  T2 Operation

The T2 timer has only two modes of operation.  The mode is controlled by ACR5.  If this bit is 0, T2 operates as a one-shot with no output, similar to the corresponding mode of T1, except that the T2 flag bit in the IFR is set when the counter goes through zero.  The other mode, selected by ACR5=1, decrements T2 each time the PB6 line goes low.  The T2 flag is set when the counter goes through zero.  T2 has associated with it only two addresses, $A008 and $A009.  Writing to $A008 loads the low order T2 latch.  Writing to $A009 loads the high order counter, transfers the low order latch to the counter, clears the T2 flag, and starts the operation.  Reading $A008 clears the T2 flag and gives the value of the low order counter byte.  Reading $A009 gives the high order counter byte.  T2's use of PB6 takes priority over any other assignment.

A reasonably versatile frequency meter can be built using the AIM-65 and a pair of NAND gates.  Use an NAND gate to invert the output of VIA pin PB7 (header pin 18).  Connect the inverted signal to one input of the second NAND gate.  Connect the signal whose frequency is to be determined (FG TTL output) to the other input of the second NAND gate.  Connect the second NAND output to PB6.  Enter the following program:

```
NEW
10 A=40960
20 POKE A+11,128+64+32
30 INPUT "PERIOD(MS)";P
35 INPUT "CTFM FREQ";FT
40 N=INT(P/0.001)-2
50 NL=INT(N-INT(N/256)*256)
60 NH=INT(N/256)
70 POKE A+8,0
80 POKE A+9,0
90 POKE A+4,NL
100 POKE A+5,NH
110 WAIT A+13,64
120 C=-(PEEK(A+8)+256*PEEK(A+9)
130 F=1000*C/P
140 PRINT F;100*(FT-F)/FT;"%"
150 GOTO 70
```

RUN the program.  Use a period of 50 msec.  Determine the range of the frequencies which can be measured.  Note that T2 cannot respond to inputs above 500 kHz.  Attach the printout showing the measurement results (P6).

<u>Question 6</u>.   Briefly describe what each statement in the above program does. Account for the differences between the CTFM reading and the computer-calculated frequency.

<u>Question 7</u>.   Suggest a method by which a period meter could be constructed. Give both a diagram and the program. Use the asynchronous waveform from figure 10-44 of the text or from figure 10-24 in Unit 10.

## 13-4  DIGITAL-TO-ANALOG CONVERTER INTERFACING

Objectives:   To connect the Signetics NE5018 digital-to-analog converter to the AIM-65 and to use the converter in a computer-controlled function generator.

A block diagram of the NE5018, taken from the Signetics literature, is shown in figure 13-2. Four sub-units are combined on the single chip: the latches and switch drivers, which are basically digital; the reference voltage generator; the multiplying DAC switches, resistors, and input buffer amplifier; and the output current-to-voltage converter. The latch section of the NE5018 will not be used in these experiments, since the 6522 VIA already performs the latching function.

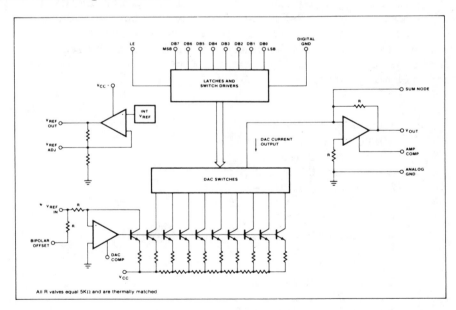

Fig. 13-2

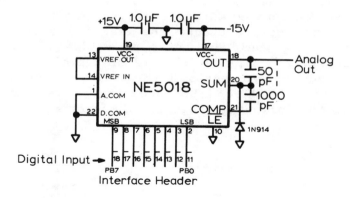

Fig. 13-3

Connect the NE5108 DAC to the 6522 VIA as shown in figure 13-3. In order to use the converter, port B must be set for output. The following program can be used to test the steady state ("DC") response of the converter:

```
NEW
10 A=40960
20 POKE A+2,255
30 INPUT "NUMBER";N
40 IF (N<0 OR N>255) THEN 30
50 POKE A,N
60 INPUT "VOLTAGE = ";V
70 GOTO 30
```

Statement 60 is used to allow the recording of the voltage on the output tape.

Run the program, and record the voltages obtained on the output for inputs of 0, 1, 2, 3, 4, 5, 10, 11, 31, 32, 63, 64, 127, 128, 253, 254, and 255.

Question 8. Using the data acquired above, estimate the offset, accuracy, linearity, and differential non-linearity of the converter. Attach the printout (P8).

A general purpose waveform generator can be constructed using only the NE5018 and the AIM-65. An integer array, A%(), contains the values representing the waveform. A%(0), the first element of the array, contains the number of elements to use. The waveform results from sending these bytes to the NE5018 DAC via port B. Unfortunately, as is often the case, the speed of the BASIC interpreter is so low that a waveform generator based upon a BASIC program would be very slow indeed. Therefore, a machine language routine is used to take the elements from the array and send them to the DAC. Timer T1 is set up in the free-running, non-output mode, and is used to control the time between presentation of the waveform values. This routine cycles repetitively through the array. In order to stop the program, the CB2 input is connected to BSI MA. After each cycle through the array, the IFR bit corresponding to CB2 is used for this function because there is a CB2 mode in which the CB2 flag is set on an active edge, but is reset only by writing IFR3=1. CB1 would not work, because the flag is reset each time port B is written. Thus the flag would be reset each time a new array element was written, and would never be high at the end of a cycle.

RESET the computer, and enter BASIC. Reserve about 256 locations by entering 3840 to the "MEMORY SIZE?" question. Press escape to enter the monitor, and enter the following program, in mnemonic entry mode:

```
<I>
0000      *=F80
0F80 20 JSR BEFE        ;get integer value
0F83 18 CLC             ;calc array address
0F84 A5 LDA 77
0F86 69 ADC #07
0F88 85 STA FE
0F8A A9 LDA #00
0F8C 65 ADC 78
0F8E 85 STA FF
0F90 A9 LDA #40         ;set T1 free-running non-output mode
0F92 8D STA A00B
0F95 A5 LDA AD          ;set period in T1
0F97 8D STA A004
0F9A A5 LDA AC
0F9C 8D STA A005
0F9F A0 LDY #01         ;get # of elements
0FA1 B1 LDA (FE),Y      ;from A%(0)
0FA3 AA TAX             ;store in X register
0FA4 AD LDA A00D        ;check T1 flag
0FA7 29 AND #40
0FA9 F0 BEQ 0FA4
0FAB AD LDA A004        ;clear T1 flag
0FAE C8 INY             ;point to next value
0FAF C8 INY
0FB0 B1 LDA (FE),Y      ;get value
0FB2 8D STA A000        ;send to DAC
0FB5 CA DEX             ;array done?
0FB6 D0 BNE 0FA4        ;no
0FB8 AD LDA A00D        ;check CB2
0FBB 29 AND #80
0FBD F0 BEQ 0F9F
0FBF 4C JMP COD1        ;return to BASIC
0FC2 (hit ESC)
```

Re-enter BASIC via the 6 key, and enter the following program:

```
NEW
10 DIM A%(100)
20 A%(0)=100:A=40960
30 POKE A+2,255
40 POKE A+12,32
50 POKE A+13,255
60 POKE 4, 128:POKE 5,14
```

```
70 FOR I=1 TO A%(0)
80 A%(I)=128+100*SIN(6.28*I/A%(0))
90 NEXT I
100 INPUT "PERIOD, MSEC";P
110 P=P*1000-2
120 IF P>32767 THEN P=P-65536
130 X=USR(P)
140 POKE A+13,255
150 GOTO 100
```

Connect the scope to the output of the DAC, and run the program. Observe the effect of periods of 2, 1, 0.2, 0.1, 0.05, 0.02, 0.01, and 0.005 ms. Measure the duration of each value of the waveform for each period.

Question 9. Explain why there is no change in the output duration when the lowest period values are used. Attach the printout (P9).

Change the FOR loop (70-90) to give a square wave.

Question 10. Show the listing of the program you use (P10). How would you change the program to give a triangular wave?

Generate a random output waveform. Remember the DAC input is restricted to 0-255.

Printout 11. Attach a listing of your program.

13-5 DEMONSTRATION OF SERVO-TYPE ANALOG-TO-DIGITAL CONVERTERS

Objectives: To construct and characterize staircase, tracking, and successive approximation servo-type analog-to-digital converters using the NE5018 DAC, a comparator, and the AIM-65.

Servo-type converters use a comparator to determine when the output of a DAC matches the unknown input voltage. The output of the ADC is the binary value necessary to produce the match. The differences among the servo ADCs are the algorithms used to find the matching value. Using a computer to implement the algorithms is somewhat like attacking roaches with a hand grenades, but the principles will be made clear. Only in fairly unusual circumstances would one actually construct an ADC in this manner.

All three types of converter will use the same hardware configuration. Connect the NE5018 DAC to the VIA as shown in figure 13-3. Figure 13-4 shows the connection of the unknown voltage and the DAC output to the comparator, and the method by which the comparator output is connected to the port A input, PA0 (header pin 1). Use the 311 comparator on the op amp job board.

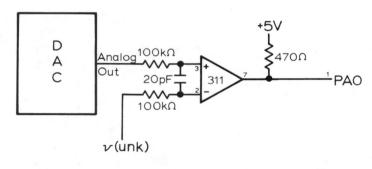

Fig. 13-4

A.  Staircase ADC

The simplest type of servo ADC is the staircase type.  This ADC simply starts the DAC at zero, and increments the digital value until the DAC output exceeds the input voltage.  The following program does this:

```
NEW
10 A=40960
15 INPUT "VOLTAGE";V
20 POKE A+2,255:POKE A+3,0
30 N=0
40 POKE A,N
50 IF (PEEK(A+1) AND 1) THEN 100
60 N=N+1
70 IF N>255 THEN PRINT "TOO HIGH":END
80 GOTO 40
100 PRINT "VALUE IS ";N
```

Run the program with input voltages of 0, 0.5, 1.0, 2.0, 5.0, 7.5, and 9.6 V. Check the linearity of the converter.

Printout 12.  Attach the printout.

To measure the speed of the converter, insert the following statements:

```
25 FOR I=1 TO 25
100 NEXT I
110 PRINT "VALUE IS";N
120 INPUT "TIME(S)";T
```

List the program.  Run the program with input voltages of 0.5, 1.0, 5.0, 7.5, and 9.5 V.  Record on the tape the time between depressing RETURN and the printing of the value.

Question 13.  What correlation exists between the input value and the conversion time?  Attach the printout (P13).

## B. Tracking ADC

Tracking ADCs use an algorithm which either increments or decrements the digital value, according to the sign of the difference between the input voltage and the DAC output voltage. Enter the following program:

```
NEW
10 A=40960
20 POKE A+2,255:POKE A+3,0
30 PRINT N
35 FOR I=1 TO 300:NEXT I
40 POKE A,N
50 D=PEEK(A+1) AND 1
60 If D=1 AND N>0 THEN N=N-1:GOTO 30
70 IF D=0 AND N<255 THEN N=N+1:GOTO 30
80 PRINT " AT LIMIT"
90 GOTO 40
```

Statement 35 is a delay to clarify the display. Turn off the printer, at least during the first runs. Observe the display while varying the input voltage in the range from 0V to 9.5V. Note that the displayed value follows or tracks the input voltages. Now stop the program. Set the voltage to 0V.

Turn the printer on. Run the program. Slowly increase the input voltage to about 1.0V. Stop the program when the value begins to oscillate.

Question 14. Why does the print-out wander between two values? Upon what does the conversion time of the tracking ADC depend? When might this type of converter be useful? Attach the printout (P14).

Question 15. The tracking ADC and staircase ADC can be implemented in hardware using an up/down counter, a DAC, and a comparator. Sketch such a circuit.

Question 16. How does the resolution of the DAC affect the resolution of the ADC which uses the DAC?

## C. Successive Approximation ADC

The successive approximation type ADC can also be implemented using the DAC and comparator connected to the AIM 65. The procedure here is to make guesses about the digital value needed by the DAC to match the input value. The method is straightforward: Set the most significant bit HI. If the DAC output is too large, set it LO again. In any event, proceed to the next least significant bit, applying the same test. For an eight bit converter, the entire process will take eight trials, regardless of the voltage input.

The following program performs the successive approximation conversion.

```
NEW
10 A=40960
15 INPUT "VOLTAGE";V
20 POKE A:2,255:POKE A:3,0
30 N=0
40 T=128
50 FOR I=1 TO 8
60 POKE A,N+T
65 PRINT "N";N;" T ";T
70 IF (PEEK(A+1) AND 1) = 0 THEN N=N+T
80 T=T/2
90 NEXT I
100 PRINT N
```

Run the program with the input voltage set to 0, 1, 2, 5, 7, and 9.5V. Also find the voltage difference between two successive digital outputs (i.e. the resolution of the converter). Determine the linear range of the converter.

Printout 17. Attach the printout.

Alter the program so that the voltage, instead of N, is output.

Printout 18. Provide listings to demonstrate that your modification works.

One problem with this type of converter is that the voltage must be constant during conversion, or difficulties ensue.
Set the input voltage at 7V. Run the program, but decrease the voltage to zero about 1 second after hitting RETURN. Repeat, starting at 0V and sweeping quickly up to 7V, again about 1 second after starting.

Question 19. Based on your observations, how steady does the input voltage have to be in order to guarantee no effect upon the conversion? Attach the printout (P19).

Remove statements 65, and add whatever statements are needed to modify the program so that the conversion time can be measured, and measure it.

Printout 20. Record your results, and document the method you used to make the measurement.

## 13-6   THE INTEGRATED CIRCUIT SAMPLE-AND-HOLD AMPLIFIER

Objectives:  To demonstrate the use of an integrated circuit sample-and-hold amplifier.

The solution to varying input voltages is to use a sample-and-hold amplifier (SHA) to sample the input and hold it until the conversion is complete. The National Semiconductor LF398 is an integrated circuit SHA. The block diagram and pin out of this device is shown in figure 13-5.

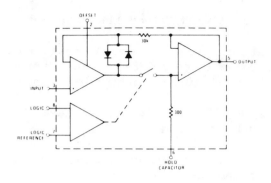

Fig. 13-5

The LF398 is connected as shown in figure 13-6 on the interface job board.

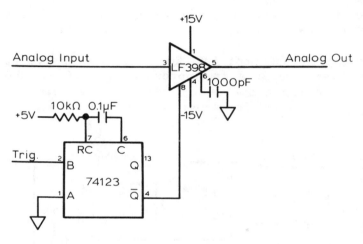

Fig. 13-6

Disconnect the LF398 output from the ADC0804 input if it has been left connected. Connect the the 398 logic input to the FG TTL output. Check to make sure that the FG output is a 0-4 V sine wave at 100 kHz, and connect it to the LF398 analog input. Connect the scope Ch1 to the FG output, Ch2 to the LF398 output, and trigger on the FG TTL output rising edge.

Observe and record the waveforms. Measure the settling time, the size of the hold offset, and droop rate (if they can be measured).

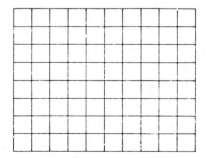

Reconnect the logic input of the LF398 to the $\overline{Q}$ output of M1 (pin 4).

<u>Question 21.</u>  How long after the hold pulse must one wait before starting conversion if a 0 to 5V 8-bit successive approximation ADC is used?  What about 10 bit and 12 bit converters?

<u>Question 22.</u>  What effect would the hold offset have on measurements?  How can this problem be overcome?

The integrated circuit sample and hold amplifier will be used throughout the remainder of this unit and unit 14.

## 13-7  INTEGRATED CIRCUIT SUCCESSIVE APPROXIMATION ADC

Objectives:  To introduce and exercise the National Semiconductor ADC0801 integrated successive approximation analog to digital converter.

The National Semiconductor ADC0804 8 bit successive approximation analog-to-digital converter converts in approximately 100 μs and can handle inputs from 0 to 5V.  Voltages above 5V or below 0V can damage the chip.  A block diagram of the chip taken from the National literature is shown in figure 13-7.

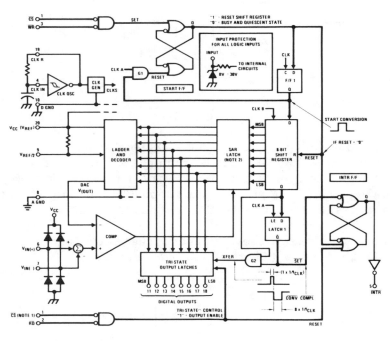

Note 1:  $\overline{CS}$ shown twice for clarity.
Note 2:  SAR - Successive Approximation Register.

Fig. 13-7

Connect the ADC0804 to the 6522 VIA through port A as shown in figure 13-8.

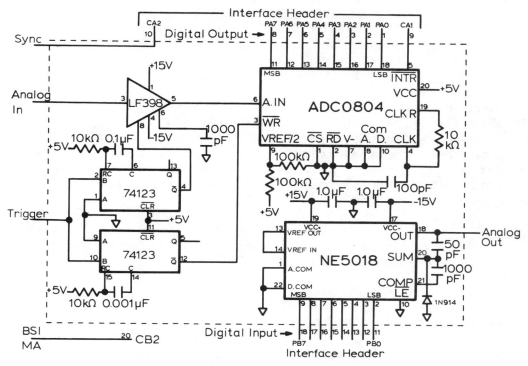

Fig. 13-8

Note that the NE5018 should be left connected, or reconnected, as shown in figure 13-8. The CHIP SELECT (CS) on the ADC0804 line is held LO, which enables the READ (RD) and WRITE (WR) inputs. To start a conversion, the WR line must go from HI to LO to HI again. When the WR line goes from HI to LO, the interrupt (INTR) line goes HI. It remains HI until the conversion is complete. Some 1 to 8 ADC clock periods after the WR line goes back HI, the conversion actually begins. The falling edge of the INTR signal can be used to load the port A register and signal the computer that a new conversion value is ready. For this reason, the INTR line is connected to the CA1 input of the VIA. Port A will be configured to latch the data on the falling edge of CA1.

The two monostables are used to start the conversion. On the rising edge of the triggering signal, which is connected to the A inputs of the monostables, M1 and M2 both start. M1 places the SHA in hold mode for the duration of the conversion. A flip flop could be used instead of M1, but would have required another chip. M2 delays for the time needed to assure that the SHA is settled, and then starts the conversion. For the moment, the trigger source is BSI job board pushbutton switch MB. Enter the following program.

```
NEW
10 A=40960:POKE A+2,255:POKE A+3,0
20 POKE A+11,1:POKE A+12,1
25 INPUT "VOLTAGE";V
30 WAIT A+13,2
40 PRINT PEEK (A+1)
50 GOTO 30
```

Connect the LF398 analog input to the VRS +1 V output and to the DMM. Verify that the LF398 output is between 0 and 5 V. Then connect the LF398 output to the ADC analog input. Run the program. Press MB to start the conversion. Test the converter by recording the input voltage and number output for a wide variety of voltages in the 0-5 V range. In the course of your experiment, determine the minimum change in voltage which gives three consecutive number outputs.

Question 23. Is the converter linear? Is it accurate to within 1 least significant bit? Use your data to support your answer. Attach the printout (P23).

13-8   ASSEMBLY LANGUAGE DATA LOGGER

Objectives:  To develop a useful data logger using the ADC0804 and the AIM-65.

It is often useful to acquire data at regular intervals and then to be able to "play them back" when convenient. It should come as no surprise that BASIC programs are often far too slow to perform this function. In this experiment an assembly language program of moderate complexity is entered and used to acquire data at a programmable sampling rate. The data acquisition program actually adds the values acquired to the values already in the array, so that averaging can be performed.

Connect the PB7 output of the interface header to the monostable trigger inputs, in place of BSI MA. RESET the computer, enter BASIC using a MEMORY SIZE of 3840. Enter monitor via ESC, and enter the following routine:

```
<I>
0000      *=F00
0F00 20 JSR BEFE      ;this is familiar code
0F03 18 CLC
0F01 A5 LDA 77
0F06 69 ADC #07
0F08 85 STA FE
0F0A A9 LDA #00
0F0C 65 ADC 78
0F0E 85 STA FF
0F10 A0 LDY #01
0F12 B1 LDA (FE),Y
0F14 AA TAX
0F15 AD LDA A001      ;clear CA flags
0F18 AD LDA A00D      ;check for sync pulse
0F1B 29 AND #01
0F1D F0 BEQ 0F18
```

```
0F1F A5 LDA AD          ;Question 1
0F21 8D STA A004
0F24 A5 LDA AC
0F26 8D STA A005
0F29 AD LDA A001        ;clear flags again
0F2C C8 INY             ;Question 2
0F2D C8 INY
0F2E AD LDA A00D        ;wait for INTR from ADC
0F31 29 AND #02
0F33 F0 BEQ 0F2E
0F35 18 CLC             ;add now to current array element
0F36 AD LDA A001
0F39 71 ADC (FE),Y
0F3B 91 STA (FE),Y
0F3D 88 DEY
0F3E A9 LDA #00
0F40 71 ADC (FE),Y
0F42 91 STA (FE),Y
0F44 C8 INY
0F45 CA DEX             ;array done?
0F46 D0 BNE 0F2C
0F48 4C JMP COD1        ;return to BASIC
0F4B (press ESC)
```

Question 24. Describe the function of the blocks of instructions marked Question 1 and Question 2 in the above listing.

The BASIC program below sets DDRA, DDRE, ACR, PCR, and then clears the IFR. Re enter BASIC via "6" and enter the program.

```
NEW
10 DIM A%(50)
20 A%(0)=50
30 A=40960
40 POKE A+3,0
50 POKE A:11,128+64+1
60 POKE A+12,4
70 POKE A:13,255
80 INPUT " SMPL TIME, MSEC";S
90 S=(S/2)*1000-2
100 IF S>32767 THEN S=S-65536
110 POKE 4,0:POKE 5,15
120 X=USR(S)
130 FOR I=1 TO A%(0)
140 PRINT A%(I);
150 NEXT
```

Question 25. Explain the function of statements 40-70 and 100-110. Why is S divided by 2 in 90?

Disconnect the ADC input from the LF398 output. Set the FG output to give a 4 V p-p triangle wave centered on 2.5 V. Use a frequency of 10 Hz. Connect the FG TTL output to VIA pin CA2 (header pin 10). Run the program. Try a sampling time of 1 msec.

Run the program twice with identical parameters. Note the reproducibility of the waveforms. Then do at least three other runs varying the input waveform and the sampling interval.

Printout 26. Attach your printout.

Determine the minimum sampling period.

Question 27. Given the conversion time of about 100 μsec, how long does the data acquisition program take, exclusive of the conversion time? Attach your pringout (P27).

The data acquisition program adds the new value to the current array value. This allows accumulation of values, all of which occur at the same time relative to the synchronizing pulse.

Verify that the addition occurs as described. Modify the program so that a variable number of waveforms are added together. Use an INPUT statement to acquire this number.

Question 28. What can go wrong when more than 256 waveforms are added together? Attach a listing of the program you use to do it. Try the program once with 500 waveforms (P28).

## 13-9  COMPLETE DATA LOGGER WITH OUTPUT

Objectives:  To implement a multi-function data logger using preprogrammed machine language routines for data acquisition, scope display and printer plotting.

In this section, the data acquisition program and the scope display program are combined with a printer/plot routine. These three machine language routines are then linked to BASIC programs to provide a versatile data logging program. The routine for data acquisition and display are similar to the ones developed above, but perform the VIA initialization in machine language, obviating the VIA POKEs. Each routine will be discussed separately.

The data acquisition routine begins at address $0E00. To use this routine from BASIC, set the USR address (bytes 4 and 5) to 0 and 14, respectively. The array into which the samples are ADDED is the first DIMensioned in the program. The array must be an integer array. The 0th element of the array contains the number of elements to use, which must be less than 128. The argument in the USR call is one-half the number of microseconds per sampling interval. This routine returns upon completion of the acquisition, and cannot be aborted once started except by the RESET button of the AIM-65.

The scope display routine begins at address $0E43. To use this routine, set the USR locations 4 and 5 to 57 and 14, respectively. As above, the data are taken from the first array, which must be an integer array. The 0th element contains the number of elements in the array to display, which must be less than 128. The argument in the USR call is the number of microseconds between displayed values. This routine displays continuously until the CB2 flag is activated. The array values should be scaled between 0 and 255 before the display routine is called.

The plot routine uses the AIM-65 plotter as a strip chart recorder. Data in the first array in memory, which must be an integer array, are plotted with 0 on the left of the tape, and 139 on the right edge. The 0th element of the array contains the number of array elements to be plotted, which must be less than 128. The array must be scaled between 0 and 139 before calling the program. The USR address bytes 1 and 5 should be set to 208 and 14, respectively, prior to calling this routine. The plot routine returns to BASIC only after completing the plot; to abort the program, RESET must be pressed.

The routines themselves are too long and complex to enter by hand. The complete listings of the programs are given in Appendix F. Consult your instructor for information on loading programs from cassette tape. Once the tape recorder has been connected, set the recorder to the beginning of the program DTLGR. Perform the following operations by entering the underlined commands. Start by RESETting the computer.

```
ROCKWELL AIM 65

<L>IN=T F=DGLTR T=1
(turn the recorder to PLAY)
(when the cursor reappears, stop the recorder)

<5>
MEMORY SIZE? 3580
WIDTH? 20
3050 BYTES FREE
AIM 65 BASIC V1.1
```

At this point, set the tape recorder to the beginning of the BASIC program called BSCDL. LOAD this program:

```
LOAD
IN=T F=BSCDL T=1
(turn recorder to PLAY)
(when cursor reappears, stop the recorder)
LIST
```

Question 29. Briefly describe what each statement or group of statements in the program does.

Connect a 100 Hz 4 V peak-to-peak triangle wave centered at 2.5 V to the analog input of the sample and hold.

Explore the operation of the program. Run the program with sample time of 1 ms and display time of 0.1 ms. Run the program at least 5 more times, using different sampling times, input wave shapes, and input amplitudes. Be careful not to exceed the limits of 0-5 V on the ADC0804.

Printout 30. Attach the printout.

Add instructions to the program which check to be sure that the sampling and display time values are reasonable.

Printout 31. List your additions and show that they work.

If instructed by the instructor to do so, save your modified program on cassette tape.

The data logger and the above routine will be used extensively in Unit 14. Leave the system connected and the program loaded if possible.

# UNIT 14.  SIGNAL PROCESSING AND NOISE REDUCTION TECHNIQUES

In this unit, a variety of signal processing and data acquisition techniques are developed.  The characteristics of random noise are presented, and methods for its elimination are demonstrated.  Periodic interference is removed by node rejection techniques.  The effect of quantization noise is shown.  Fourier transform techniques are used to analyze waveforms and to demonstrate aliasing.  A brief introduction to digital filtering is given, and in the process, a multichannel averager, a multichannel analyzer, and a boxcar integrator are constructed.  Finally, least squares polynomial smoothing is introduced as a numerical signal-to-noise enhancement technique.

This unit is primarily software-related, and most of the experiments are based upon the data logger developed in Unit 13.

## 14-1  USE OF DTLGR AND THE BDLGR SKELETON

Objectives:  To become familiar with the interaction between the DTLGR machine language routines and the BDLGR BASIC language support package.

The complete data acquisition circuit is shown in figure 14-1.  This circuit provides an 8-bit, 0-5V analog input, an 8-bit, 0-1 V analog output, and a variety of timing signals and logic-level inputs.  This circuit, with a few slight modifications, is used throughout this unit.

The machine language routines in DTLGR (DaTa LoGgeR) have been discussed in the previous unit.  In this unit, these routines are combined with a "skeleton" of BASIC language subroutines, stored as program BDLGR (Basic Data LoGgeR).  With the exception of the Fourier transforms and digital filtering experiments, all the experiments in this unit are performed by making relatively minor additions to the BDLGR routines.  In this section, the objective is to simply verify that the data acquisition circuit and the software package function properly.

Wire the circuit in figure 14-1.  In the rest of this unit, the entire circuit of figure 14-1 is treated as a block, with inputs SYNC, Trigger, and ANALOG IN, and outputs ANALOG OUT and PB7.  The digital in and digital out lines are never altered during the experiments.  Make no changes to the wiring of the Data acquisition job board after the circuit of figure 14-1 has been wired unless you are specifically requested to do so.  Signals entering and leaving the Data acquisition block are identified in figure 14-1.  Connect the scope Ch 1 to the ANALOG OUTPUT, and wire Ch 2 to the output of the LF398 sample-and-hold chip (pin 5).

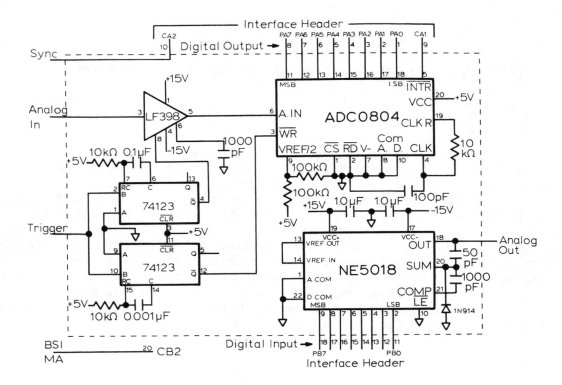

Fig. 14-1

Turn on the computer, and load DTLGR and BDLGR from tape as follows:

```
            ROCKWELL AIM-65
            <L>IN=T F=DTLGR T=1
            (set recorder to beginning of DTLGR and "PLAY" the tape)
            (when cursor reappears, stop the recorder)

            <5>
            MEMORY SIZE? 3580
            WIDTH? 20
              3050 BYTES FREE
              AIM 65 BASIC V1.1
              LOAD
              IN=T F=BDLGR T=1
            (set recorder to beginning of BDLGR and PLAY again)
```

Connect the analog input to a 2 V p-p 20 Hz sine wave centered at 2 V. BE SURE TO MEASURE THE INPUT WAVEFORM ON THE SCOPE BEFORE CONNECTING IT TO THE DATA ACQUISITION SYSTEM!!! Connect the TTL output of the function generator to the SYNC input of the data acquisition unit (CA2, interface header pin 10).

The BDLGR program consists of two sections. The first, and shorter section is the "driver", which is changed in each experiment. The driver for this section is:

```
10 REM BDLGR
20 REM DATA LOGGER
30 REM SKELETON
40 DIM A%(100)
50 A%(0)=100
60 GOSUB 5000
70 GOSUB 6000
80 CLEAR:GOTO 10
```

When the program is run, the subroutine at line 5000 of BDLGR acquires the data at the rate (in milliseconds per point) specified by the user. The number of cycles selected determines the number of times the data acquisition routine is run; each time through the routine the new values obtained and added to the previous sum. The subroutine at 6000 finds an appropriate scale factor for the data, and displays them on the scope screen, by means of the analog output. Pressing PB1 on the BSI board terminates the scope display and records the data on the printer tape.

Run the program. Specify a sampling time of 1 ms. Use 1, 2, and 4 cycles. Adjust the scope Ch 1 so that the analog display is stable and accurately represents the data in the array. Note that the display program injects a single point lower than any data at the beginning of the trace to aid in synchronizing the scope display. Sketch the resulting display.

Question 1. Are there any changes in the data between runs? Does this agree with your expectations?

14-2  MULTICHANNEL AVERAGING

Objectives:  To use the data acquisition system as a multichannel averager and to determine its signal-to-noise enhancement capabilities.

The data logger as constructed in experiment 14-1 is a multichannel analyzer. Each element in the A% array represents the sum of the values for several analog-to-digital conversions on the waveform at a fixed time from the beginning of data acquisition. The start of acquisition is determined by the SYNC pulse, which is the LO-to-HI transition of the TTL output of the function generator.

Install the Reference job board in the breadboard frame. Disconnect the ADC0804 analog input from the LF398 output. Connect the FG output to the signal input of the noise generator circuit. Adjust the FG offset and the noise amplitude so that the output of the noise generator is a 2 V p-p 20 Hz sine wave centered at 2 V, to which 2 V p-p noise has been added. Be certain that the output of the noise generator does not exceed 5 V! After checking that the output of the noise generator is betwen 0 and 5V, connect it to the data acquisition system analog input.

Run the program. Select a sample time of 1 ms and 1 cycle of data acquisition. Repeat for 2, 4, 16, 64, and 256 cycles.

Question 2.    Estimate the signal-to-noise ratio for each of the runs above. Compare the improvement with the theoretically expected square-root-of-n improvement expected for random noise.

Question 3.    Given that the integer array A% can only take numbers in the range of -32768 to +32767, what is probably wrong with the output for the 256 cycle run?

## 14-3 NOISE REDUCTION BY AVERAGING

Objective:  To investigate further the enhancement of signal-to-noise ratio by averaging.

A more quantitative analysis of the noise reduction properties of averaging can be performed by using the following program. Note that only the statements in the range of 0 to 200 are shown; the subroutines at 5000 and 6000 should remain unchanged. Therefore, carefully delete the statements in the previous program except for the subroutines. Do not type NEW!

```
10 REM NR
20 REM NOISE
30 REM REDUCTION
40 DIM A%(100)
50 A%(0)=100
60 GOSUB 5000
70 AV=0
80 FOR I=1 TO 100
90 AV=AV+A%(I)
100 NEXT I
110 AV=AV/100:SD=0
120 FOR I=1 TO 100
130 SD=SD+(A%(I)-AV)*(A%(I)-AV)
140 NEXT I
150 PRINT "AVERAGE:";AV/CY
160 PRINT "STD DEV:;SQR(SD/99)/CY
180 CLEAR:GOTO 10
```

Disconnect the noise generator output from the data acquisition system analog input. Reduce the sine-wave amplitude to 0V, but leave the 2 V p-p noise and the 2 V offset. Check that the noise generator output is really 2 V p-p noise centered at 2 V, and reconnect the noise generator output to the data acquisition system analog input.

Run the program, using a 1 ms sample interval and 1, 2, 4, 16, and 256 cycles. Repeat each run.

Question 4. What does the program do? Explain the "/CY" in statements 150 and 160.

Question 5. How does the standard deviation change with the number of cycles? How does this compare with the expected behavior? Account for any changes in the average and standard deviation between runs.

Save this program. It is used in the next section.


## 14-4 REJECTION OF PERIODIC NOISE

Objectives: To demonstrate the rejection of periodic interference by appropriate choice of sampling interval.

The program "NR" in the previous experiment, together with the BDLGR subroutines, is used again in this experiment, with the following additions:

```
55 INPUT "FREQ:";F
165 GOSUB 6000
```

Adjust the function generator output to obtain a 2 V p-p 20 Hz sine wave centered at 2 V. Reduce the noise amplitude to zero. The frequency must be accurately set in this experiment, so use the CTFM, in the period mode if necessary. Without altering the offset in any way, reduce the sine wave amplitude to zero.
Run the program, using a 1 ms sampling interval and 1 cycle. Attach the printout giving the average value and the standard deviation (P5).
Increase the sine wave amplitude to 2 V p-p. Run the program again with a sampling time of 1 ms and 1 cycle. Use input frequencies of 20, 21, 24, 25, 29, 30, 50, 60, and 100 Hz. Attach the printout (P6).

Question 6. For which frequencies is the noise rejected? On the basis of the above data only, what conclusion can you draw about relationship needed between the frequency of the interference and the sampling interval for rejection of the interference?

The above experiment is artifical in that there is synchronization between the TTL function generator output, which starts the conversion process, and the interferring signal. Normally, this is not the case. To simulate asynchronous interference, disconnect the SYNC input, CA2, from the function generator TTL output, and connect it instead to MB on the BSI board.
Repeat the above program, except this time use MB to start the data acquisition. Because accidental coincidence may occur between the pushing of MB and the start of the wave, repeat each run three times. Use 20, 21, 25, 29, and 30 Hz. Attach the printout (P7).

Question 7. Under what conditions is the interference rejected? What range of "average" values would you expect if a 55 Hz signal was input to the data acquisition system?  Account for the magnitude of the standard deviation observed.

Question 8. An alternative to summing points over the period of the interferring noise might be to sample only twice per cycle of the noise.  Would this be effective?  Would sampling once per cycle work?  Why?

## 14-5  QUANTIZATION NOISE

Objectives:  To examine the nature of quantization noise, which is noise due to the finite number of levels in the analog-to-digital conversion process, and to demonstrate a method by which it can be reduced.

The fact that the analog-to-digital converter used in these experiments has a only 256 different possible output values leads to a distortion of any digitized waveform simply because intermediate values cannot be represented. This distortion is called QUANTIZATION NOISE.  Quantization noise differs from random noise and periodic interferences in that neither averaging nor variations in sampling interval will remove it.  The characteristics and methods for reducing quantization noise are the subject of this experiment.

Connect the function generator to the noise generator signal input and adjust the generators so that the output is a 2 V p-p, 1 Hz triangular wave centered at 2 V, with no noise added.  Connect this signal to the analog input of the data acquisiton system.  Connect the SYNC input of the data acquisition system (6522 pin CA2, pin 10 on the header) to the function generator TTL output.  The trigger input of the data acquisition system comes from 6522 pin PB7, header pin 18.

Remove all BASIC statements except for the subroutines, and enter the following program (which is identical to BDLGR except for the REM statements):

```
10 REM QUANT NOISE
20 REM DEMONSTRATION
30 REM PROGRAM
40 DIM A%(100)
50 A%(0)=100
60 GOSUB 5000
70 GOSUB 6000
80 CLEAR:GOTO 10
```

Run the program.  Select 1, 4, and 16 cycles.  Use a sampling interval of 0.5 ms.  Attach the printour (P9)

Question 9. Calculate the number of points expected at each quantization level of the analog-to-digital converter, and compare to the number observed.

Question 10. Some smoothing of the edges of the transitions between levels will probably be observed in the runs with 4 and 16 cycles.  Why does this occur?

Question 11.   Estimate the signal-to-noise ratio in the traces obtained in the program just run above. Compare your estimate to the theoretical value, and account for any discrepancies.

Increase the noise amplitude to about 30 mV.  Run the program, again with 0.5 ms sample interval, using 1, 4, 16, and 64 cycles.  Then reduce the triangular wave amplitude to zero, and run the program, again with 1 cycle and 1 ms sampling interval.  After the run, restore the triangular wave amplitude to its previous value.

Rerun the program two more times, first with 60 mV noise amplitude, then with 120 mV noise.  Label all three printouts (P12).

Question 12.   From the traces in which the triangular wave amplitude is zero, determine the number of quantization levels spanned by the noise.

Question 13.   Estimate the signal-to-noise ratio and effective resolution shown in each of the traces obtained above.  Formulate a strategy for the most efficient method for the elimination of quantization noise, and the improvement of resolution, based upon your results.  Compare your method with the theoretical optimal strategy, and explain any differences.


14-6  MULTICHANNEL ANALYZER

Objectives:   To introduce the concept of multichannel analysis by constructing a simple multichannel analyzer, and to determine the amplitude spectrum of the random noise source for a number of different settings.

A multichannel analyzer is a device which records the number of times each amplitude occurs in the signal.  "Occurances" may be discrete events, with the amplitude of each event recorded, or they may be the result of sampling a continuous waveform.  The latter process is discussed here, although the former is the more common use of the multichannel analyser.

The sampling operation will be performed by a subroutine in machine language which is embedded in DTLGR, but which previously has not been used. This routine uses timer 1 in the single pulse output mode to trigger the data acquisiton system a variable number of microseconds after the routine is entered.  It will be extensively used in the boxcar integrator of experiment 14-7.  The USR call which enters this routine contains as its parameter the number of microseconds  to be delayed.  The value returned by the USR function is the result of the conversion.

The multichannel analyzer program uses the display subroutine of the BASIC skeleton, but not the data acquisition section.  Nevertheless, delete only lines numbered less than 5000 before entering the following program:

```
10  REM MULTI
20  REM CHANNEL
30  REM ANALYZER
40  DIM A%(100)
50  A%(0)=100
60  POKE 4, 169:POKE 5,14
70  INPUT "SAMPLES";S
80  FOR I=1 TO 100
90  C=USR(10)
100 IF C>99 THEN PRINT "ERROR":GOTO 120
110 A%(C+1)=A%(C+1)+1
120 NEXT I
130 GOSUB 6000
140 CLEAR:GOTO 10
```

Set the noise generator output for about 1 V p-p noise centered at about 1 V.

Run the program. Use 10, 100, 1000, and 10,000 samples. Reduce the noise amplitude to about 0.5 V p-p and repeat, again with 10, 100, 1000, and 10,000 samples. Label the printout P14.

Question 14.   Compare the widths of the curves obtained with the p-p amplitudes. Are the results reasonable?

Question 15.   How nearly gaussian are the curves? On the basis of the curves, how good a random noise source is the MM5837 noise source?

14-7  BOXCAR INTEGRATOR

Objectives:   To build and study a scanning boxcar integrator, and to apply it to noise reduction problems.

The construction of a boxcar integrator uses the routines in DTLGR and BDLGR. In this experiment, the delayed sample routine, discussed in the previous section, is used to acquire the data. This routine uses timer 1 in the output one-shot mode to start an external event, and to start the conversion at the end of the one-shot period. Thus samples can be taken any number (up to 65535) of microseconds after the beginning of any event which can be started by the timer 1 output. Timer 1 uses PB7 as its output; this pin is also connected to the DAC, but the DAC will not be in use when the data are being collected.

As a representative signal, the charging voltage on the capacitor of a 555 timer will be used. Disconnect the noise generator output from the data acquisition system analog input (LF398 input). Wire the circuit shown in figure 14-2 on the Signal conditioning job board.

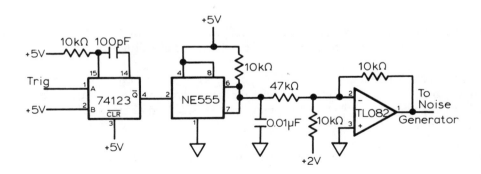

Fig. 14-2

Temporarily connect the trigger input of the monostable to the FG TTL output, so that the circuit is triggered at about 100 Hz. This circuit produces an output which should be an exponentially falling waveform of about 1 V p-p amplitude with an offset about 2.5 V below common. Connect this output to the input of the noise generator, with the noise amplitude set to zero. The noise generator output should be an exponentially rising waveform of about 1 V p-p amplitude offset about 2.5 V above common. The output of the noise generator should not be connected to the analog input of the data acquisition system until the voltage levels have been verified.

Enter the following program which uses the BDLGR output subroutine:

```
10 REM BOXCAR
20 REM INTEGRATOR
30 REM BX1
40 DIM A%(100)
50 A%(0)=100
60 INPUT "N";N
70 INPUT "T";T
80 POKE 4,169:POKE 5,14
90 FOR I=1 TO 100
100 AV=0
110 FOR J=1 TO N
120 AV=AV+USR(I*T)
130 NEXT J
140 A%(I)=AV/N
150 NEXT I
160 GOSUB 6000
170 CLEAR:GOTO 10
```

In this program, N is the number of times the program samples at each delay time, and T is the increment in delay time, in microseconds.

Connect the trigger of the monostable on the signal conditioning job board to PB7 (header pin 18) instead of the TTL FG output.

Run the program, with a variety of different N and T values. It may help to use T=1 and N=500 first, to be sure the program is functioning properly. Trigger on Ch 1 (the ANALOG OUTPUT). Now increase the noise amplitude to

272

about 3 V p-p. The signal should be difficult to see in all the noise. Run the program with T=2 and N= 1, 4, 16, 64, 256, and 1024. Label the printouts P17.

Question 16.   Comment on the observed signal-to-noise ratio enhancement and compare with your theoretical expectations.

Rewrite the program so that N scans are made through the data, each pass taking only one conversion at each delay, instead of the above situation in which one scan is made of N conversions at each delay. If time permits, enter and run this program.

Question 17.   What difference would there be in the two methods?

Question 18.   What advantages and disadvantages does the boxcar integrator have in comparison to the multichannel averager?

14-8   FOURIER TRANSFORM AND ALIASING

Objectives:   To introduce and demonstrate the fast Fourier transform (FFT), and to demonstrate the aliasing effect on under sampled waveforms.

The actual FFT routine, together with a data acquisition routine, is recorded on tape as file FFT. This routine also uses the DTLGR data acquisition and printer/plotter routines. The routines in BDLGR will not fit in the 4K memory of the AIM-65 at the same time as the FFT routine. Load the FFT routine and the DTLGR routines as follows:

```
            Reset the computer
                ROCKWELL AIM-65
            <L>IN=T F=DTLGR T=1
            (set recorder to beginning of DTLGR and PLAY)
            (when cursor reappears, stop the recorder)

            <5>
            MEMORY SIZE? 3580
            WIDTH? 20
               3050 BYTES FREE
                AIM BASIC V1.1
            LOAD
            IN=T F=FFT T=1
            (Set recorder to beginning of FFT and PLAY)
```

The program consists of three sections. Statements 0 - 999 constitute the driver routines, which handle data acquisition and the loading of the arrays. The subroutine at 1400 handles the formatting and scaling of arrays for printing. If Z=0 when 1400 is called, only the FR array is plotted. If Z<>0, then the square root of the sum of the squared elements of FR and FI are plotted. This gives a power spectrum output.

The FFT program is a translation into BASIC of the FORTRAN program found in "Digital Signal Analysis" by S. D. Stearns, p 226. This routine is packed very tightly, and is difficult to read. We do not propose to explain the operation of the FFT routine in detail in this experiment, but rather, the utility of the FFT is demonstrated by example.

The output of the FFT program consists of the power spectrum of the input waveform. The frequency increment between points is equal to the sampling frequency divided by the number of points in the transform array. The first element in the output array represents zero frequency; the second point in the array represents the first positive frequency; the third point the second positive frequency, etc. However, the power spectrum includes both positive and negative frequencies. The first negative frequency is the last point in the array, with successive points working toward the center of the array. The 64th point represents either the positive or the negative frequency. Since the data to be transformed in this experiment are all real numbers, the amplitudes at positive and at negative frequencies will be identical.

Connect the FG output to the noise generator, and adjust the generators so that a 200 Hz, 2 V p-p sine-wave centered at 2 V emerges from the output of the noise generator. For now, decrease the noise amplitude to zero. Check for a safe signal amplitude (0-5 V). Connect the noise generator output to the analog input of the data acquisition system (figure 14-1). The TTL FG output should be reconnected to SYNC. The TRIGGER input connects to PB7. As before, monitor the analog output on scope Ch 1, and the LF398 output on scope Ch 2.

Run the program with input frequencies of 200, 50, 100, 400, 600, 800, and 1200 Hz. Label the printout P19.

Question 19. What is the sampling frequency used in this program? (Be careful the USR argument is less than one-half the period of the T1 timer output square-wave.) Do your results support the Nyquist theory, which states that signals are represented properly if sampled at greater than twice their frequency. Is aliasing observed? If so, make a correspondence table that matches input frequencies and observed frequencies.

To reduce the dc component of waveform, which tends to reduce the vertical resolution of the outputs, the average value of the input data can be subtracted from each value before performing the transform.

Insert code between statements 30 and 40 to find the average of the A%(I) values. Change the code in line 50 to subtract the average from each value before assigning it to FR(I). DO NOT insert any spaces between variables and operators, as they are not required and take up space, which is at a premium in this program. Rerun the program, again with an input frequency of 200 Hz. Label the printout P20.

Question 20. Comment on the differences between the power spectra obtained in printouts 19 and 20.

Run the program again, using 50 Hz square and triangle waves. (Printout P21).

Question 21. Are the results as anticipated? Is there any evidence of aliasing?

Unwanted frequency components can be removed from the input waveform by transforming the waveform into the frequency domain, zeroing the undesirable frequencies, and performing an inverse transform. The inverse transform can be accomplished by running the FFT program on the frequency domain data after inverting the sign of the data in the FI array, which represent the imaginary frequency elements. A very crude filter can be accomodated in the limited memory of the AIM-65 by simply retaining the amplitude at one frequency and zeroing all other elements. Enter the following code:

```
90 INPUT"IX";IX
100 Z=FR(IX):S=FI(IX):J=FR(130-IX):L=FI(130-IX)
110 FORI=1TON:FR(I)=0:FI(I)=0:NEXT
120 FR(IX)=Z:FR(130-IX)=J:FI(IX)=-S:FI(130-IX)=-L
130 GOSUB1000
140 Z=0:GOSUB1400
150 END
```

In this routine, IX is the number of the element in the positive frequency power spectrum to be retained.

Run the program with the input set to a 2 V p-p 50 Hz sine-wave centered at 2 V to which is added 2 V p-p noise. As always, check the output of the noise generator before connecting it to the data acquisition system. From previous printouts, determine number of the point in the power spectrum which contains most of the 50 Hz signal power, and enter that value as IX. (Attach printout 22).

Question 22. Do the results that you obtain on the inverse transform agree with your expectations? Are there distortions? Compare the result with the multichannel averaging operations done previously.


14-9 SMOOTHING OPERATIONS*

Objectives: To demonstrate, on synthetic data, the degree of noise reduction and signal distortion which results from the use of least squares smoothing functions.

Smoothing operations are basically noise reduction techniques which remove high frequency fluctuations by approximating sections of the data by polymonials, and then using the value of the polynomial at the center point of the section as a "better" approximation of the data at that point. Savitsky and Golay (Anal. Chem. 36(8), 1627 and also Steinier et. al., Anal Chem 44(11), 1906) published sets of coefficients to be used in smoothing data according to various polynominal fitting functions. The operative equation is:

------------------

*Based on an experiment written by T.A. Nieman.

$$Y(j)* = \sum_{i=-M}^{M} \frac{C(i)Y(j+i)}{n}$$

where Y(j) and Y(j)* are the original and smoothed data points, respectively; the C(i) factors are the weighting coefficients, N is a normalizing coefficient, and —m to m is the range of data points around the central value which are used in the smooth.

The parameters available to the experimenter include the form of the smooth, which determines the C(i), the width of the smooth, which determines m, and the number of times the smooth is applied. These types of smoothing operations cause some distortion in the signal; the severity of the distortion is a complex function of the above parameters.

Load the SMOTH program. Use the same procedure as for FFT, including the loading of the DTLGR program.

The program consists of four parts. Statements 10 through 200 interact with the experimenter, and create gaussian waveform of the width and signal-to-noise ratio desired. The RND(X) function, not previously used, produces a new random number between 0 and 1 if X is positive; produces the previous random number if X is 0; and starts a new series of random numbers if X is negative. The section at 300 through 390 calculates the smoothing coefficients, and prints them. SMOTH is a skeleton, to which 340 and 360 must be added before smoothing can take place. Statements at 340 and 360 are used to enter the specific smoothing function. The section at 500 through 590 performs the actual smoothing operation. The code at 1000 scales the output array and calls the DTLGR print routine.

For each of the following functions, enter the corresponding 340 and 360 statements, and run the program. Use a full width at half maximum (FWHM) of 8, and a signal-to-noise ratio of about 8. Smooth the data using smooths of 1, 2, and 3 times the FWHM of the synthetic. Label the prints P23a through P23d.

Function 1 (Simple Moving Average):

$$Cs = \frac{1}{2M + 1}$$

```
340 D=2*M+1
360 N=1
```

Function 2 (Triangular Moving Average):

$$Cs = \frac{(M+1) - |s|}{(M+1)^2}$$

```
340 D=(M+1)*(M+1)
360 N=M+1-ABS(S)
```

Function 3 (Quadratic-Cubic):

$$Cs = \frac{3(3m^2 + 3m - 1 - 5S^2)}{(2m+3)(2m+1)(2m-1)}$$

```
340 D=(2*M+3)*(2*M+1)*(2*M-1)
360 N=3*(3*M*M+3*M-1-5*S*S)
```

Function 4 (Quartic-Quintic):

$$Cs = \frac{15}{4} \frac{(15m^4+30m^3-35m^2-50m+12)-35(2m^2+2m-3)S^2+63S^4}{(3m+5)(2m+3)(2m+1)(2m-1)(2m-3)}$$

```
340 D=(2*M+5)*(2*M+3)*(2*M+1)*(2*M-1)*(2*M-3)
360 N=1,25*(15*M*M*M*M+30*M*M*M-35*M*M-50*M+12)
361 N=N-35*(2*M*M+2*M-3)*S*S+63*S*S*S*S
```

<u>Question 23</u>.  For each function, graph the coefficients as a function of S, as obtained from the data tape.  Comment on the relative merit of the four functions with respect to noise reduction, distortion, and speed.  Also comment on the effect of changing the smooth width to FWHM ratio.  Identify, if possible, the optimum value of the ratio for each function.

Enter the two following functions, and run the program again.  Be sure to delete statement 361 from the previous program.  Use FWHM of 8, a signal-to-noise ratio of 10, and a smooth width of 8.  Label these printouts P24a and P24b.

Mystery Function 1:

$$Cs = \frac{3s}{(2m+1)(m+1)(m)}$$

```
340 D=(2*M+1)*(M+1)*M
360 N=3*S
```

Mystery Function 2:

$$Cs = \frac{30(3s^2 - m(m+1))}{(2m+3)(2m+1)(2m-1)(m+1)(m)}$$

```
340 D=(2*M+3)*(2*M+1)*(2*M-1)*(M+1)*M
360 N=30*(3*S*S-M*(M+1))
```

Question 24. What do the mystery functions do? What utility might these functions have?

Question 25. What happens to the first and last $(N-1)/2$ points when an N point smooth is used? Do new values get assigned to these points?

## APPENDIX A.  LABORATORY STATION

The laboratory station consists of standard test equipment, available from commercial suppliers, a breadboard frame and power supply, pre-configured patchboards called "job boards" and miscellaneous cables, leads, connectors and loose parts.  A description of all necessary parts and equipment is included in this Appendix.

1.  Standard Test Equipment
    a)  Oscilloscope
        Specifications:
        1.  >5 MHz bandwidth
        2.  Dual trace
        3.  XY
        4.  Triggered sweep
        5.  Z-axis modulation desirable, but not required
        Suitable Models:
            Tektronix 2213
            Ballantine 1032A or 1010A
            Gould OS255
            Heathkit IO4450
            Hameg HM203
    b)  Function Generator
        Specifications:
        1.  1 Hz - 1 MHz frequency range
        2.  Sine, square, triangular wave outputs
        3.  TTL output
        4.  Variable dc offset
        Suitable Models:
            Wavetek 180                  Kron-Hite 1200A
            Global Specialties 2001      Simpson 420A
    c)  Digital Multimeter
        Specifications:
        1.  3 1/2 digits
        2.  0 - 100 V, 0 - 2 A, 0 - 2 M$\Omega$
        Suitable Models:
            Keithly 130                  Weston 6100
            Beckman 310
            Fluke 8022B
            Simpson 461 or 463
    d)  Counter/Timer/Frequency Meter (CTFM)
        Specifications:
        1.  >6 digit
        2.  DC to 10 MHz
        3.  Frequency, period, period average, frequency ratio, time interval, events count
        4.  Time Base, decade selectable 10 ms to 10 s.

The experiments utilize the Intersil 7226 EV/Kit, 10 MHz Universal Counter Evaluation Kit, an inexpensive CTFM.  We have found it desirable to mount feet on the PC board and to install a plexiglass cover stood off from the board to accomodate the switches.  The CTFM is powered from an auxiliary output on the power supply (+5 V is required).  A parts list and assembly instructions are given below.

CTFM Parts List:

| Item | Qty | Description | Source |
|------|-----|-------------|--------|
| 1. Intersil 7226A EV/Kit | 1 | Counter kit and documentation | Jameco |
| 2. BNC chassis mount, females | 2 | Signal input connectors | |
| 3. Banana jacks | 3 | Store, measure and reset output jacks | |
| 4. Toggle switch FTE01/JMT-121 | 1 | Run/Hold switch | Jameco |
| 5. Pushbutton switch, 275-1547 | 1 | Reset switch | Radio Shack |
| 6. Feet and Screws | 4 | Any size rubber | |
| 7. Plexiglass sheet 6"x9"x1/8" | 1 | Cover | |
| 8. Connector | 1 | 4 Pin TRW Cinch Jones Male | |
| 9. Bolts and Nuts, 6-32 | 4 | | |
| 10. Spacers, 1/2" | 4 | | |

CTFM Assembly Instructions:

1. Build counter kit as described in Intersil instruction.
2. Solder in Reset switch.
3. Solder in input jumpers A and B across patch space.
4. Drill 2 holes in PC board and mount Run/Hold switch between +5 V and pin 39 counter chip. Cut off unused switch lead. Solder and epoxy in place.
5. Drill holes at corners of PC board and mount feet.
6. Drill holes in plexiglass sheet as shown below.

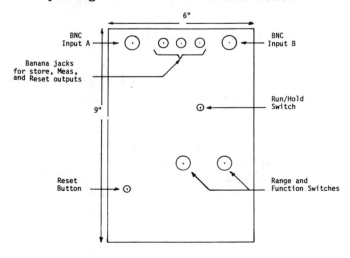

7. Wire one BNC connector as input A and the other as input B. Solder wires directly to tab on PC board (see Intersil instructions).
8. Wire banana jacks to store, measure and reset on tab (see Intersil instructions).
9. Make up a 2-conductor, strain-relieved cable as power cable ($\simeq$ 2 feet). Solder one end to +5 V and common tab connections on PC board. Attach 4-pin, male Cinch Jones connector to other end.
10. Install plexiglass cover with standoffs and install knobs and switches (see sketch below).

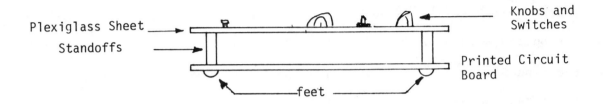

e) Digital Panel Meter (Unit 9 only)

An inexpensive digital panel meter (DPM) kit is used in Unit 9 to study dual slope analog-to-digital conversion. We have modified the Intersil 7106 EV/kit for this experiment to include a power On/Off switch, an increased range (0-2 V) and two additional test points.

DPM Parts List:

| Item | Qty | Description | Source |
|---|---|---|---|
| 1. Intersil 7106 EV/Kit | 1 | LCD Panel Meter | Jameco |
| 2. SPDT switch | 1 | On/Off switch | |
| 3. 9V battery | 1 | power source | |
| 4. 100 k$\Omega$, 5% resistor | 1 | | |
| 5. T-44 pins | 2 | Test pins | |

DPM Assembly Instructions:

1. Build DPM kit as described by Intersil.
2. Drill 2 holes for On/Off switch below R1 and through $V^+$ trace on back of circuit board. Epoxy and solder switch leads, one to each end of broken $V^+$ trace.
3. Solder in place 100 k$\Omega$ resistor on the back side (foil side) of PC board between GND and end of R5 not directly connected to input.
4. Insert TP6 on front side by drilling a hole by left leg of C4, inserting a T-44 pin, soldering and epoxying in place (see figure below).
5. Insert TP 7 on right side of R2 (see figure below).

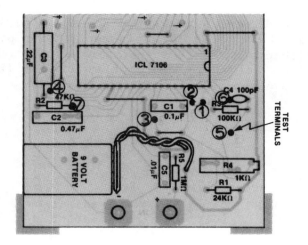

f)  Logic Probe
    Specifications:
    1.  Speed, 1 MHz
    2.  Level detection and pulse
    Suitable Models:
        Global Specialties LP2
        Radio Shack 22-300
        Jameco kit LPK-1
g)  Curve tracer (Unit 7 only)

For the Unit 7 experiments one characteristics curve tracer should be available for the laboratory.

    Suitable Models:
    1.  Heath kit IT-3121
    2.  Leader LTC-905
h)  Microcomputer (Units 12, 13 and 14 only)

The microcomputer used in Units 12-14 is the Rockwell AIM-65 with 4 kbytes RAM and BASIC interpreter in ROM with power supply and cassette tape recorder.  An enclosure for the supply and computer is also desirable.  A list of accessory suppliers for the AIM-65 is available from Rockwell.  With some modifications to programs and instructions, APPLE or PET computers can also be used.

2.  Breadboard frame and Power Supply

a)  Breadboard frame

The experiments are greatly facilitated by the breadboard frame.  In conjunction with the power supply, the frame provides +5 V, +15 V, -15 V and common connections to breadboard cards plugged into its connectors.  Signals are bussed to each card from 5 BNC connectors and 5 pairs of banana jacks. This flexible input/output connector system solves a major problem of solderless breadboarding, that of providing easy connection for function generators, DMM´s, scopes, frequency meters and other test instruments.  In addition the breadboard frame provides a solid support for patchwiring to and on the job boards.  The frame is easily made from a plywood base and a printed circuit board for the bussed signals and power. A tab on each breadboard socket mates with a PC board socket soldered to the printed circuit board. Five job boards are accommodated in the system. A parts list and the assembly instructions are given below. A description of the tabs attached to each breadboard is given in Appendix B.

## Breadboard Frame Parts List:

| Item | Qty | Description | Source |
|------|-----|-------------|--------|
| 1. Plywood base, 16"x12"x1/2" | 1 | | Lumber yard |
| 2. Printed Circuit Board | 1 | Power and signal distribution | See Note below |
| 3. Feet | 4 | Rubber (any size) | |
| 4. Barrier strip, 7-140 TRW-Cinch with Y type lugs | 1 | | Newark #28F705 or Pioneer |
| 5. Card extender socket, 50-245-30 TRW-Cinch | 5 | | Newark #29F1415 |
| 6. Banana jacks | 10 | 5 Red 5 Black | |
| 7. BNC Connectors, Case mount female | 5 | | |
| 8. Screws 6-32, 1 1/4" | 7 | | |
| 9. Screws 6-32, 1/2" | 4 | | |
| 10. Nuts 6-32 | 11 | | |
| 11. Standoffs, 7/16" tall | 7 | | |
| 12. Capacitors, 1μF, 25V tantalum | 10 | | |
| 13. Capacitors, 6.8 μF, 15 V tantalum | 5 | | |

Note: We are working to establish a convenient source for the printed circuit board required for the breadboard frame. For further information write "Printed Circuit Boards," P.O. Box 27315, Lansing, Michigan 48909.

## Breadboard frame Assembly Instructions:

1. Obtain plywood base and drill 7 mounting holes (for 6-32 screws) for PC board as shown below.

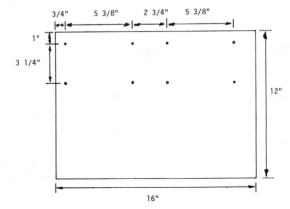

2. Install 4 feet at corners of plywood base.
3. Obtain PC board and mount all banana jacks and BNC connectors as shown below (black banana jacks to common line, red to signal lines). Connect center of BNC's to PC board with 22 gauge solid wire.

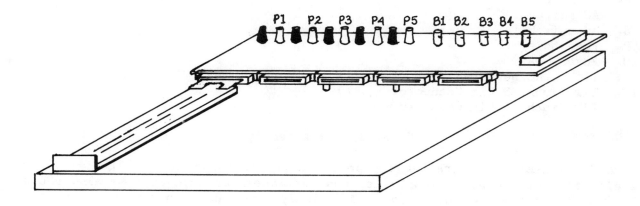

4. Mount terminal strip on right side of PC board with 6-32, 1/2" screws and nuts. Solder 4 Y-type lugs to power bus underneath. Connect 3 unused terminal contacts to 3 pins on right most edge connector for external transformer inputs.
5. Solder the 5 24 pin extender sockets to the PC board card edge.
6. Solder the 15 decoupling capacitors on the PC board as shown on the board.
7. Attach 7/16" high spacers between PC board and plywood and attach PC board to plywood base with 7, 6-32, 1 1/4" screws and nuts. See diagram below.

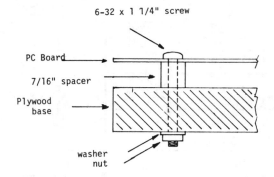

b) Power supply

The power supply consists of a commercial open frame triple supply and an external transformer placed in an enclosure with switches, fuses, lights, and auxiliary outlets.

Power Supply Parts List:

| Item | Qty | Description | Source |
|------|-----|-------------|--------|
| 1. CBB-75 W Open frame power supply | 1 | + 5 V at 6A<br>± 15 V at 1.5 A | Power One |

| | | | | |
|---|---|---|---|---|
| 2. | Bud box Cu-7125 | 1 | Enclosure | Newark 91F900 |
| 3. | Transformer, 12.6 V rms with center tap | 1 | | Newark 05F1412 |
| 4. | Line cord | 1 | | |
| 5. | Strain relief | 1 | | |
| 6. | AC Lights | 2 | | |
| 7. | AC switch 3A | 1 | | |
| 8. | AC switch 1A | 1 | | |
| 9. | Fuse holders | 4 | | |
| 10. | Fuses, 1A | 4 | | |
| 11. | 115 V, AC sockets | 2 | | |
| 12. | TRW Cinch, 3 Pin Connector | 1 | | |
| 13. | TRW Cinch, 4 Pin Connector | 3 | | |

Power Supply Assembly:

The schematic diagram of the power supply is shown in the figure below. A pictorial of the power supply box and the labels we have used is also shown. In the laboratory we have found it highly desirable to connect auxiliary signal sources, such as the function generator, into the switched ac outlets. This practice assures that these sources will be turned off when power to the job boards is turned off. The auxiliary dc outputs can power external instruments that do not have an internal power source, such as the counter/timer/frequency meter and the logic probe, the external transformer is used with the power supply job board. It is connected to the right most frame card slot by a cable between the terminal strip on the frame and the TRW cinch connector on the power supply box.

Power Supply Schematic Diagram

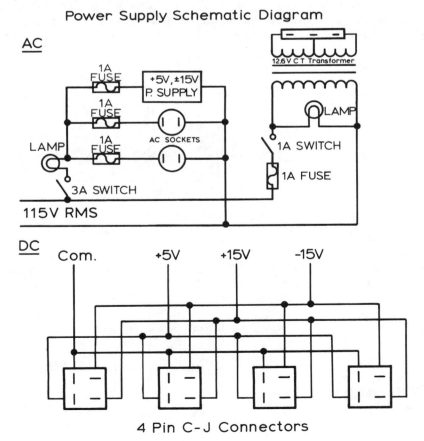

4 Pin C-J Connectors

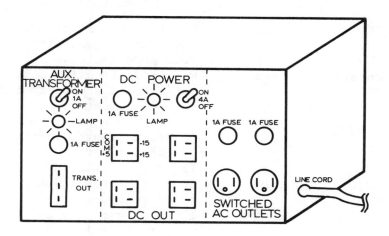

3.  Job Boards

     The job boards used in the experimental system consist of a breadboard
socket (SK-10 socket, E and L Instruments, Derby, Conn or equivalent), a
printed circuit tab, an aluminum face plate, and prewired components.
Assembly instructions and component layouts are given in Appendix B.  The job
boards fall into the following 3 categories:

     1)  Prewired Circuits that perform specific preconnected functions, such
as binary indication or the voltage reference function.  On these boards all
connections except for input/output connections are permanent and the student
should be cautioned not to alter this wiring.
     2)  IC job boards which contain integrated circuits and components and
have only power connections prewired.  The student does all the wiring on
these job boards, except for the power connections.
     3)  Utility job boards which contain only the power bus.  These are used
for testing loose components and special "single use" circuits.
     A list of all job boards and a classification of them is given below.
Job boards 1-7 in the list below are used throughout the 14 units.  Thus one
of these is needed per station.  Job boards 8-13 are used only infrequently
(one or two units).  Thus a few of these can be made up and shared between
several stations.

Job Board List:

| Name | Units used in | Category and Description |
|------|---------------|--------------------------|
| 1. Reference (REF) | all | Prewired circuits; precision time base, noise generator, voltage reference source. |
| 2. Binary switch and indicator (BSI) | all | Prewired circuits; 8 logic level switches, 2 momentary switches, 10 LED indicators. |
| 3. Op amp | most | IC's; 8 op amps, 1 comparator with prewired inputs, 1 quad analog switch, 2 resistor arrays. |
| 4. Basic gates | most | IC's; basic TTL logic circuits. |
| 5. Flip-Flop | most | IC's; flip-flops and a 4-bit shift register. |
| 6. Signal Conditioning | several | IC's; op amp, comparator, timer, monostable multivibrators, driver, relay, photo FET. |
| 7. Utility | most | Utility |
| 8. Counter | 4,10 | IC's and prewired display; decade counters, latch and 1-decade display (prewired). |
| 9. Power supply | 3 | IC's and components; Regulator, power diodes, zener diodes, resistors, capacitors. |
| 10. Advanced Analog Circuits | 9 | IC's; multiplier, active filter, phase-locked loop and regulator, op amps and trim pots, lamp. |
| 11. MSI gates | 10,11 | IC's; magnitude comparator, adder, decoder, multiplexer, arithmetic logic unit. |
| 12. Advanced gates | 11 | IC's; CMOS NOR, tristate buffer, Schmitt trigger, buffer/driver, open collector NAND, low power Schottky NAND. |
| 13. Interface | 13,14 | |

4. Cables, Connectors and Components

a) Cables and Connectors
The following list of cables and connectors are needed for all the experiments.

| Item | Qty per Station |
|------|-----------------|
| 1. Cables, BNC male connectors, both ends, $\approx$ 3 feet | 5 |
| 2. Cables, banana plugs both ends | 2 (1 red, 1 black) |
| 3. Alligator clips | 2 |

b) Parts needed
The following list is compiled by Unit.

| Unit | Item | Qty per Station |
|------|------|-----------------|
| 1 | 9 V battery | 1 |
| 1 | Unknown resistor | 1 |
| | Various resistors, 1%, 5%, 10%: 47$\Omega$, 150$\Omega$, 220$\Omega$, 300$\Omega$, 1k$\Omega$(2), 10k$\Omega$, 12k$\Omega$, 15k$\Omega$, 100k$\Omega$, 220k$\Omega$, 1M$\Omega$, 10M$\Omega$ | 1 each except where noted |
| 2 | Resistors, 5%, 39$\Omega$, 270$\Omega$, 1k$\Omega$ 10k$\Omega$, 100k$\Omega$ | 1 ea. |
| | Capacitors: 0.01$\mu$F, 0.001$\mu$F unknown RC filter | 1 ea. |
| 3 | Resistors, 5%, 10$\Omega$, 1k$\Omega$, 1.2k$\Omega$, 2k$\Omega$, 2.2k$\Omega$, | 1 ea. |
| | signal diode | 1 |
| | power diode | 1 |
| | zener diode, 5.0 V | 1 |
| 4 | Resistors, 5%, 100$\Omega$, 150$\Omega$, 220$\Omega$ | 1 ea. |
| | Photovoltaic cell | 1 |
| | Thermistor | 1 |
| | Mercury thermometer | 1 for several stations |
| | Styrofoam cup, hot water, ice | 1 |
| | Strain gauge | 1 for several stations |
| | Photocell, cadmium sulfide | 1 |
| | Phototransistor | 1 |
| | Optointerrupter | 1 |
| | Variable density filter | optional |
| 5 | Resistors, 5%, 47$\Omega$, 100$\Omega$ | 1 ea. |
| | Capacitors: 1$\mu$F, 10% | 2 |
| | Signal diode | 1 |
| | Unknown capacitor | 1 |
| 6 | Resistors, 5%, 1k$\Omega$(2), 10k$\Omega$(2), 100k$\Omega$(3), 125k$\Omega$, 250k$\Omega$, 500k$\Omega$, 1M$\Omega$, 10M$\Omega$ | 1 ea. except as noted |
| | Capacitors: 0.01$\mu$F | 3 |
| 7 | Transistor, 2N2222 | 1 |
| | FET, 2N5459 | 1 |
| | Matched FET pair, 2N3958 | 1 |
| | SCR, 5-10A, 100-200 V | 1 |
| | Triac, SK2582 | 1 |
| | Resistors, 10%: 7.5$\Omega$(5W) | 1 |
| | Resistors, 5%: 56$\Omega$, 100$\Omega$, 470$\Omega$, 680$\Omega$, 1k$\Omega$, 4.7k$\Omega$, 10k$\Omega$, 22k$\Omega$, 47k$\Omega$, 100k$\Omega$(5), 220k$\Omega$ | 1 ea. except as noted |

Capacitors:  47pF, 68pF, 0.1μF, 1μF                    1 ea.
Lamp, CM53                                              1
Phototransistor, FPT-100 (Radio Shack)                 1
Potentiometer, 100kΩ                                   1
Optically isolated, zero crossing switch,              1
   Elec-trol SA-8

8     Resistors, 1%, 10kΩ, 100kΩ                       4 ea.
      Resistors, 5%:  1k , 5kΩ, 1MΩ(2)                 1 ea. except as noted

      Capacitors:  47pF, 470pF, 0.047μF, 1μF(1)        2 ea. except as noted
      Signal diode                                     4
      Photodiode                                       1

9     Resistors, 5%:  150Ω, 220Ω, 470Ω, 1kΩ, 4.7kΩ,   1 ea. except as noted
      10kΩ(2), 15kΩ

      IBM card, punched                                1

      Integrated circuit, DIP package                  1

10    None needed
11    Resistors, 5%:  47Ω, 1kΩ(2), 4.7kΩ, 10kΩ,       1 ea. except as noted
      22kΩ, 33kΩ

      Capacitors:  0.01μF                              3
      LED                                              1

12    None needed

## APPENDIX B.  JOB BOARD LAYOUTS

This appendix includes construction details for and pictorial diagrams of all job boards used in the experiments.  The breadboard sockets used to make the job boards are SK-10 sockets available from E and L Instruments, Derby, CT, or an equivalent.  A printed circuit tab plugs into one end of the SK-10 socket and mates with a PC socket on the breadboard frame.  This arrangement automatically powers each job board and allows input/output connections to be made via BNC connectors and banana jacks on the frame.  Each SK-10 socket also has an aluminum face plate which is used with the REF and BSI job boards for mounting switches and potentiometers and with the others for support while inserting and removing job boards from the frame.

1.  Tab and Face Plate

Parts List:

| Item | Qty per job board | Description | Source |
|---|---|---|---|
| 1. SK-10 Socket | 1 | Breadboard socket | E & L Instruments |
| 2. PC Tab | 1 | Tab for power and I/0 connections | See Note page 282 |
| 3. T-44 | 14 (20 for power supply job board) | | |
| 4. Face plate | 1 | 7 1/4" x 2 3/16" x 1/16" aluminum for all job boards except REF and BSI (see next section for these) | |
| 5. Panhead screws, 6-32 x 1/4" | 4 | | |
| 6-32 x 3/4" | 2 | | |
| 6. Nuts and insulating washers | 2 | | |

Assembly Instructions:

The following instructions apply to all job boards <u>except the</u> <u>REF</u> <u>and</u> <u>BSI</u> <u>job boards</u> which are covered in the next section.

1.  Obtain a connector tab PC board.
2.  Use a shear to cut the connector tab to shape.  For all job boards <u>except</u> the power supply job board, cut off the 3 pairs of holes at the bottom as shown below.

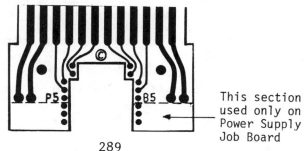

This section used only on Power Supply Job Board

289

3. Insert the 14 T-44 pins (20 pins for the power supply job board tab) and solder in place. Cut pins to a length of 1/4". The connector tab is now complete.

4. Tap all 6 holes of the SK-10 breadboard socket with a 6-32 tap.

5. Make a face plate as shown below out of 1/16" aluminum.

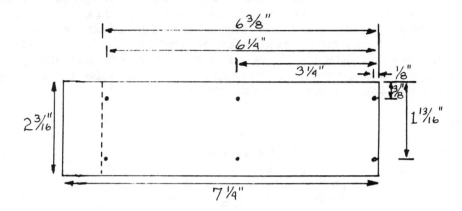

6. Connect the face plate, SK-10 and connector tab together with the 6,6-32 screws as shown below.

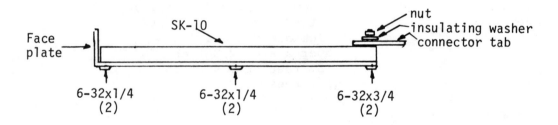

## 2. REF and BSI Job Boards

The REF and BSI job boards are more complex than the others because they contain prewired circuits and have switches and potentiometers mounted on the aluminum face plate. Thus separate instructions are given here for these boards.

### a) REF Job Board

Parts List:

| | Item | Qty | Description | Source |
|---|---|---|---|---|
| 1. | SK-10 socket | 1 | Breadboard socket | E & L Instruments |
| 2. | Connector tab | 1 | Printed circuit board | See Note page 282 |
| 3. | T-44 pins | 14 | | |
| 4. | Face Plate | 1 | 8 1/4" x 2 3/16" x 1/16" aluminum | |
| 5. | Panhead screws 6-32, 1/4" | 4 | | |

|     |                                          |           |                              |                                                                                    |
| --- | ---------------------------------------- | --------- | ---------------------------- | ---------------------------------------------------------------------------------- |
|     | 6-32, 3/4"                               | 2         |                              |                                                                                    |
| 6.  | Nuts and insu-<br>lating washers,<br>6-32 | 2         |                              |                                                                                    |
| 7.  | 1 MHz Crystal                            | 1         |                              | Jameco #CYIA                                                                        |
| 8.  | RDD 104                                  | 2         | 4 decade CMOS<br>frequency dividers | LSI Computer<br>Systems, Inc.<br>1235 Walt Whitman Rd.<br>Melville, NY  11747 |
| 9.  | 8 pin Dip switch                         | 1         |                              |                                                                                    |
| 10. | 7414                                     | 1         | Hex Schmitt inverter         |                                                                                    |
| 11. | TL082                                    | 1         | Dual op amp                  |                                                                                    |
| 12. | TL084                                    | 1         | Quad op amp                  |                                                                                    |
| 13. | MM5837                                   | 1         | Digital Noise Source         |                                                                                    |
| 14. | LH0070                                   | 1         | Precision Reference          |                                                                                    |
| 15. | Trimpot, 10kΩ                            | 1         | Noise generator adjust       | Jameco 63P103                                                                      |
| 16. | Potentiometers,<br>10k , ten-turn,<br>chassis mount | 2 | VRS output adjust          | Clarostat 73JA                                                                     |
| 17. | Resistors, 5%                            |           |                              |                                                                                    |
|     | 10kΩ                                     | 6         |                              |                                                                                    |
|     | 100kΩ                                    | 4         |                              |                                                                                    |
|     | 150kΩ                                    | 1         |                              |                                                                                    |
|     | 200kΩ                                    | 2         |                              |                                                                                    |
|     | 22MΩ                                     | 1         |                              |                                                                                    |
| 18. | Resistors, 1%                            |           |                              |                                                                                    |
|     | 10kΩ                                     | 1         |                              |                                                                                    |
|     | 90kΩ                                     | 1         |                              |                                                                                    |
| 19. | Capacitors,<br>ceramic                   |           |                              |                                                                                    |
|     | 39pF                                     | 1         |                              |                                                                                    |
|     | 100pF                                    | 5         |                              |                                                                                    |
|     | 0.001μF                                  | 1         |                              |                                                                                    |
|     | 0.2μF                                    | 1         |                              |                                                                                    |
| 20. | Solid wire,<br>22 gauge                  | as needed |                              |                                                                                    |

### Tab and Face Plate Assembly

1. Assemble the connector tab as described in the preceeding section of this appendix (cut off the 3 pairs of holes at the bottom).
2. Tap all 6 holes of the SK-10 breadboard socket with a 6-32 tap.
3. Make a face plate out of 1/16" aluminum.

4. Connect the face plate, SK-10 and connector tab together with the 6,6-32 screws as before.

5. Mount the two, 10-turn potentiometers on the front of the aluminum plate as shown below.

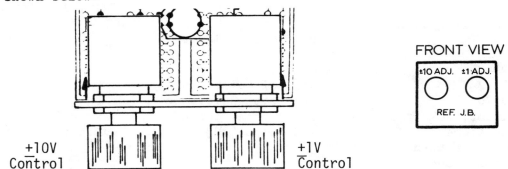

+10V
Control

+1V
Control

FRONT VIEW

±10 ADJ.    ±1 ADJ.

REF. J.B.

6. Install the integrated circuits and components and wire the frequency divider, noise generator, and voltage reference source circuits as shown in the schematic diagrams and job board layout on the next page. These permanent wires should all be as short as feasible and should lie nearly flat on the job board to discourage their removal.

7. Solder 22 gauge wires to the 10-turn potentiometers and connect them to the appropriate circuit points as shown in the VRS schematic.

8. Label the various outputs as shown on the job board layout diagram.

9. Plug the REF job board into the frame and test for appropriate voltages on the power bus. Test the frequency divider by connecting a CTFM to the 1 MHz, ÷ N, and ÷ NM outputs in turn. Test for various positions of the DIP switch (see Unit 9). Test the noise generator by connecting a scope to the noise generator output v(out). Adjust the noise generator trim pot and observe that the noise amplitude varies. Test the VRS by connecting a DMM to the various outputs in turn. Verify that the fixed outputs (+10 V, -10 V, +1 V, -1 V) give the appropriate voltages ±2% and that the variable outputs (±10 V and ±1 V) give appropriate adjustable outputs. This completes the REF job board construction and testing.

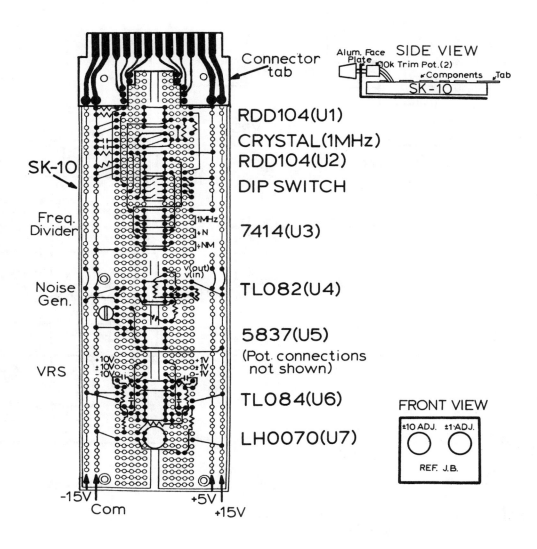

Ref. Job Board
(Component Layout)

Connector tab

SK-10

RDD104(U1)

CRYSTAL(1MHz)

RDD104(U2)

DIP SWITCH

Freq. Divider

1MHz
÷N
÷NM

7414(U3)

Noise Gen.

v(out)
v(in)

TL082(U4)

5837(U5)
(Pot. connections not shown)

VRS

+10V
±10V
-10V

+1V
±1V
-1V

TL084(U6)

LH0070(U7)

-15V
Com

+5V
+15V

SIDE VIEW

Alum. Face Plate
10k Trim Pot.(2)
Components    Tab
SK-10

FRONT VIEW

±10 ADJ.   ±1 ADJ.

REF. J.B.

293

## FREQ. DIVIDER

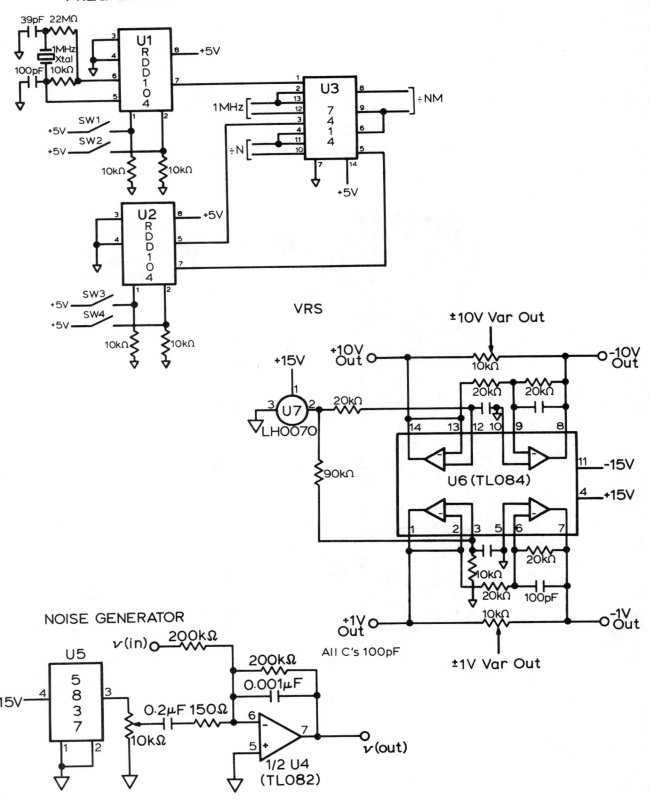

## VRS

## NOISE GENERATOR

b)  BSI Job Board

Parts List:

| | Item | Qty | Description | Source |
|---|---|---|---|---|
| 1. | SK-10 socket | 1 | Breadboard socket | E & L Instruments |
| 2. | Connector tab | 1 | PC board | See Note page 282 |
| 3. | T-44 pins | 14 | | |
| 4. | Face plate | 1 | 8 1/4" x 2 3/16" x 1/16" aluminum | |
| 5. | SPDT, 2 position switches | 8 | Binary source switches A-H | Jameco FTE01 JMT-121 |
| 6. | SPDT, momentary switches | 2 | Momentary switches MA,MB | |
| 7. | Panhead screws | | | |
| | 6-32 x 1/4" | 4 | | |
| | 6-32 x 3/4" | 2 | | |
| 8. | Nuts & insulating washers | 2 | | |
| 9. | 4050 B | 2 | CMOS driver | |
| 10. | 7404 | 4 | Hex inverter | |
| 11. | Resistors, 5% | | | |
| | 1k $\Omega$ | 10 | | |
| | 100k $\Omega$ | 10 | | |
| 12. | Miniature LEDS | 10 | Binary indicators | Jameco XC 209R |
| 13. | Solid wire, 22 gauge | as needed | | |

Tab and Face Plate Assembly

1.  Assemble the connector tab as before.
2.  Tap all 6 holes of the SK-10 socket with a 6-32 tap.
3.  Make a face plate out of 1/16" aluminum as was done with the REF job board except that 10 holes are to be drilled in the folded section for mounting switches.
4.  Connect the face plate, SK-10 and connector tab together with 6,6-32 screws as before.
5.  Mount the 10 switches on the front of the aluminum plate as shown in the layout on the next pages.
6.  Install the IC's and components and wire the circuits shown on the schematic diagram and job board layout. Again make permanent wires as short and as flat on the board as feasible. Wire the indicators as shown in the table.
7.  Solder 22 gauge wires to the 10 switches and connect them as shown in the circuit diagram and table on the next pages.
8.  Label the various outputs as shown on the job board layout diagram.
9.  Test the sources and indicators by patchwiring each binary source output (SA-SH, MA, MB) to an LED input (LA-LJ). Toggle the switches and verify correct circuit action.

# BSI Job Board

## SOURCES

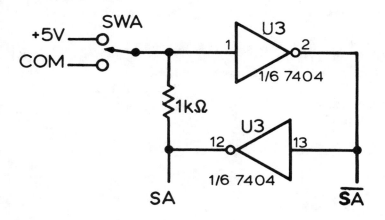

| SW# | IC# | PIN#'s |
|-----|-----|--------|
| SA | U3 | 1,2,13,12 |
| SB | U3 | 3,4,11,10 |
| SC | U3 | 5,6,9,8 |
| SD | U4 | 1,2,13,12 |
| SE | U4 | 3,4,11,10 |
| SF | U4 | 5,6,9,8 |
| SG | U5 | 1,2,13,12 |
| SH | U5 | 3,4,11,10 |
| MA | U5 | 5,6,9,8 |
| MB | U6 | 1,2,13,12 |

## INDICATORS

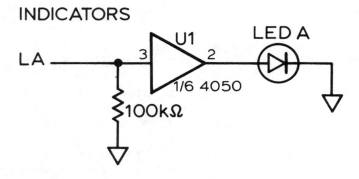

| LIGHT # | IC# | PIN#'s |
|---------|-----|--------|
| LA | U1 | 3,2 |
| LB | U1 | 5,4 |
| LC | U1 | 7,6 |
| LD | U2 | 3,2 |
| LE | U2 | 5,4 |
| LF | U2 | 7,6 |
| LG | U1 | 14,15 |
| LH | U1 | 11,12 |
| LI | U2 | 14,15 |
| LJ | U2 | 11,12 |

# BSI Job Board

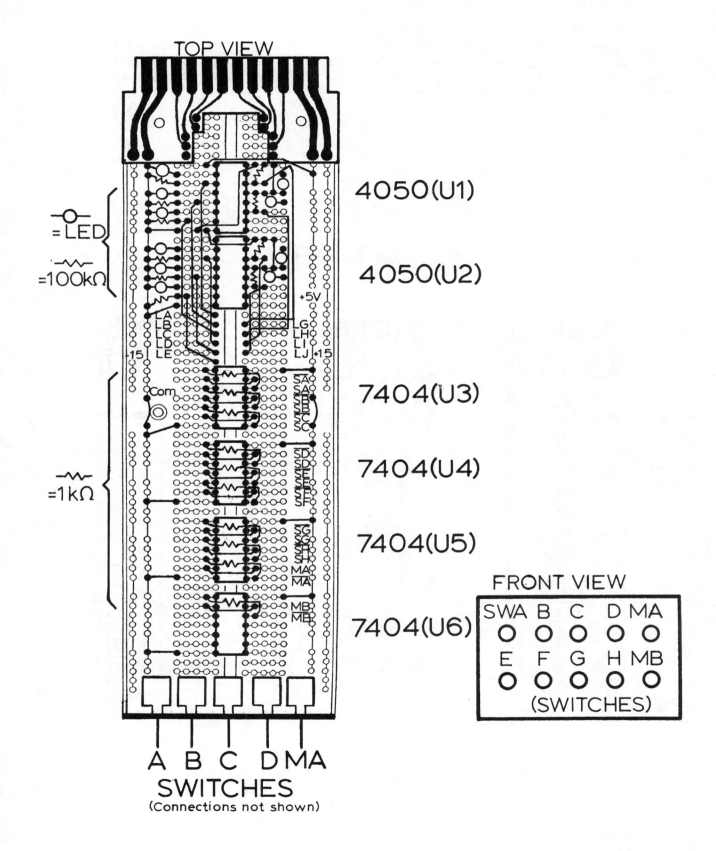

TOP VIEW

= LED

=100kΩ

4050(U1)

4050(U2)

LA
LB
LC
LD
LE

+5V

LG
LH
LI
LJ

-15

+15

Com

SA
SA
SB
SB
SC
SC

7404(U3)

=1kΩ

SD
SD
SE
SE
SF
SF

7404(U4)

SG
SG
SH
SH
MA
MA

7404(U5)

MB
MB

7404(U6)

A B C D MA

**SWITCHES**
(Connections not shown)

FRONT VIEW

| SWA | B | C | D | MA |
|-----|---|---|---|-----|
| O | O | O | O | O |
| E | F | G | H | MB |
| O | O | O | O | O |

(SWITCHES)

## 3. Other Job Boards Descriptions

The next few pages give pictorial diagrams and parts lists of all the remaining job boards in the system. The power supply job board differs from the others in that the connector tab has 3 extra pairs of contacts that bring the external transformer secondary out to the job board when it is plugged into the right-most frame position.

Op-amp Job Board    Basic Gates Job Board    Flip-flop Job Board

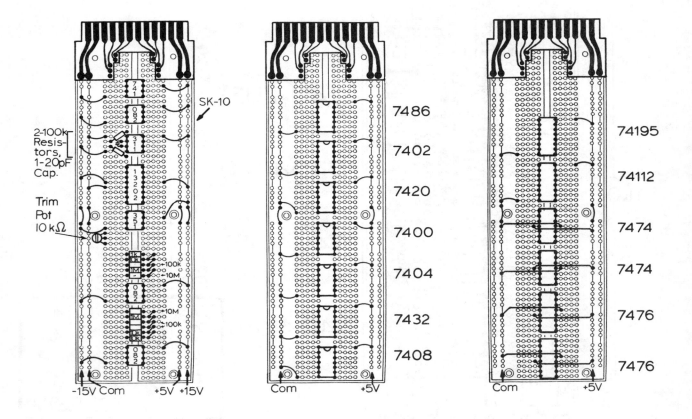

## Counter Job Board

Connector tab

LED Display

Resistors, 1kΩ

7447

7475

7490

74190

SK-10

Com      +5V

## Signal Conditioning Job Board

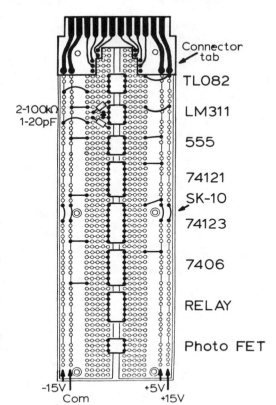

Connector tab

TL082

LM311

555

74121

SK-10

74123

7406

RELAY

Photo FET

2-100kΩ
1-20pF

-15V        +5V
   Com      +15V

## Power Supply Job Board

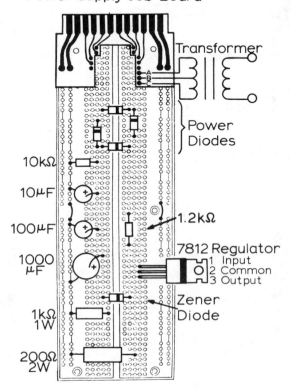

Transformer

Power Diodes

10kΩ

10μF

100μF

1.2kΩ

1000 μF

7812 Regulator
1 Input
2 Common
3 Output

Zener Diode

1kΩ
1W

200Ω
2W

300

## Advanced Analog Job Board

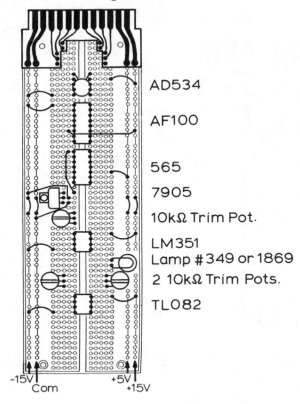

AD534

AF100

565

7905

10kΩ Trim Pot.

LM351
Lamp #349 or 1869

2 10kΩ Trim Pots.

TL082

-15V
Com
+5V
+15V

## MSI Gates Job Board

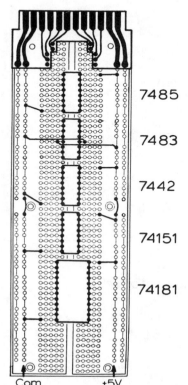

7485

7483

7442

74151

74181

Com
+5V

## Advanced Gates Job Board

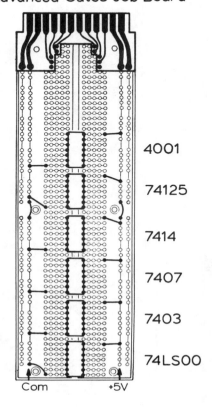

4001

74125

7414

7407

7403

74LS00

Com
+5V

## Utility Job Board

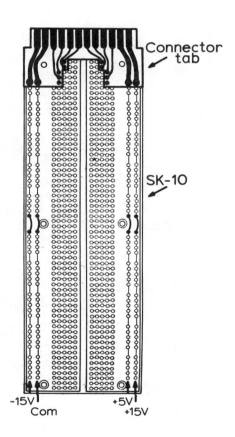

Connector tab

SK-10

-15V
Com
+5V
+15V

## Interface

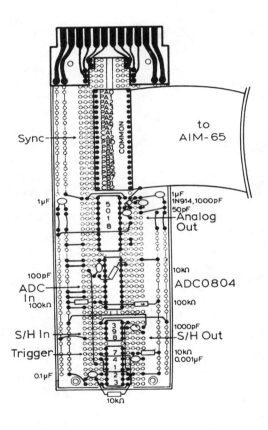

to AIM-65

Sync

1µF

1µF
1N914, 1000pF

50pF

Analog Out

100pF

10kΩ

ADC0804

ADC In

100kΩ

100kΩ

1000pF

S/H In

S/H Out

Trigger

10kΩ
0.001µF

0.1µF

10kΩ

## Parts Lists:

### (excluding connector tab and face plate)

| Job Board | Item | Qty | Source |
|---|---|---|---|
| Op amp | LF351 op amp | 1 | Jameco |
| | μA741 op amp | 1 | Jameco |
| | TL082 dual op amp | 3 | Jameco |
| | LM311 Comparator | 1 | Jameco |
| | LF 13202 Quad analog switch | 1 | Pioneer |
| | 10kΩ trim pot | 1 | Jameco 63P103 |
| | Resistors, 1% | | |
| | 1kΩ | 1 | |
| | 10kΩ | 3 | |
| | 100kΩ | 5 | |
| | 1MΩ | 2 | |
| | 10MΩ | 2 | |
| | Capacitor, 20 pF | 1 | |
| | 22 gauge wire | as needed | |
| Basic Gates | 7486, 7402 | 1 each | |
| | 7420, 7404, 7432 | | |
| | 7408 | | |
| | 22 gauge wire | as needed | |
| Flip-Flop | 74195, 74112 | 1 each | |
| | 7474, 7476 | 2 each | |
| | 22 gauge wire | as needed | |
| Counter | 7 segment display | 1 | Jameco Man 74 Red C.C. |
| | 7447, 7475, | 1 each | |
| | 7490, 74190 | | |
| | Resistors, 5% | | |
| | 1kΩ | 7 | |
| | 22 gauge wire | as needed | |
| Signal conditioning | LM 311 comparator | 1 | |
| | TL082 | 1 | |
| | NE555 timer | 1 | |
| | 74121, 74123, 7406 | 1 | |
| | SPDT Relay, PC Mount | 1 | ITT Mz-5HG |
| | H11F2 Photo FET | 1 | GE |
| | Resistors, 5% 100kΩ | 2 | |
| | Capacitor, 20 pF | 1 | |
| | 22 gauge wire | as needed | |
| Power Supply | Power diodes | 4 | |
| | Zener diode, 5 V (1N4733) | 1 | |
| | Regulator, 12 V (7812 or LM340T-12) | 1 | |
| | Resistors | | |
| | 200Ω, 2W | 1 | |
| | 1kΩ, 1W | 1 | |
| | 1.2kΩ, 1/4W | 1 | |
| | 10kΩ, 1/4W | 1 | |

|  | Capacitors, electrolytic, | | |
|---|---|---|---|
|  | 10μF | 1 | |
|  | 100μF | 1 | |
|  | 1000μF | 1 | |
|  | 22 gauge wire | as needed | |
| Advanced Analog | AD534 | 1 | Analog Devices |
| Circuits | AF100 | 1 | Jameco |
|  | PLL565 | 1 | Jameco |
|  | Regulator, −5 V | 1 | |
|  | (7905 or LM320T-5) | | |
|  | LM351 | 1 | |
|  | TL082 | 1 | |
|  | 10kΩ trim pots | 3 | |
|  | Lamp, #349 or 1869 | 1 | |
|  | 22 gauge wire | as needed | |
| MSI Gates | 7485, 7483, 7442, 7451 | 1 each | |
|  | 74181 | | |
|  | 22 gauge wire | as needed | |
| Advanced Gates | 7403, 7407 | 1 each | |
|  | 7414, 74125 | | |
|  | 74LS00 | 1 | Jameco |
|  | 4001, CMOS | 1 | Jameco |
|  | 22 gauge wire | as needed | |
| Utility | 22 gauge wire | as needed | |
| Interface | NE5018 DAC | 1 | Jameco |
|  | ADC0804 | 1 | Jameco |
|  | LF 398, sample-and-hold | 1 | Jameco |
|  | 74123 | 1 | |
|  | Resistors, 1% | | |
|  | 100kΩ | 2 | |
|  | Resistors, 5% | | |
|  | 10kΩ | 3 | |
|  | Capacitors, ceramic | | |
|  | 50 pF | 1 | |
|  | 100 pF | 1 | |
|  | 1000 pF | 1 | |
|  | 0.001μF | 1 | |
|  | 0.01 μF | 1 | |
|  | 1 μF | 2 | |
|  | Capacitors, polystyrene | | |
|  | 1000 pF | 1 | |
|  | Diode, 1N914 | 1 | |
|  | 40 conductor ribbon | 2 feet | |
|  | cable | | |
|  | 40 pin DIP header | 1 | |

## 4. Interface Job Board Assignments

A 40-conductor ribbon cable terminated in a 40-pin header connects from the AIM-65 J1 application connector to the interface job board (see Unit 13). This brings the AIM-65 parallel port (VIA) contacts out to the job board for patch wiring. The connections and pin assignments are given in the table below.

| VIA pin name | 40-pin header Pin Number | AIM-65 J1 Application Connector |
|---|---|---|
| PA 0 | 1 | 14 |
| PA 1 | 2 | 4 |
| PA 2 | 3 | 3 |
| PA 3 | 4 | 2 |
| PA 4 | 5 | 5 |
| PA 5 | 6 | 6 |
| PA 6 | 7 | 7 |
| PA 7 | 8 | 8 |
| CA 1 | 9 | 20 |
| CA 2 | 10 | 21 |
| PB 0 | 11 | 9 |
| PB 1 | 12 | 10 |
| PB 2 | 13 | 11 |
| PB 3 | 14 | 12 |
| PB 4 | 15 | 13 |
| PB 5 | 16 | 16 |
| PB 6 | 17 | 17 |
| PB 7 | 18 | 15 |
| CB 1 | 19 | 18 |
| CB 2 | 20 | 19 |
| Common | 21-40 | 1 |

Note: Application connector J1 also connects to the cassette tape recorder. See the AIM-65 user's guide for details.

# APPENDIX C.  IC AND SEMICONDUCTOR LEAD ASSIGNMENTS

Included in this appendix are the pin assignments of the most common series 74 TTL integrated circuits, lead assignments for diodes and transistors, and standard op amp pin assignments.

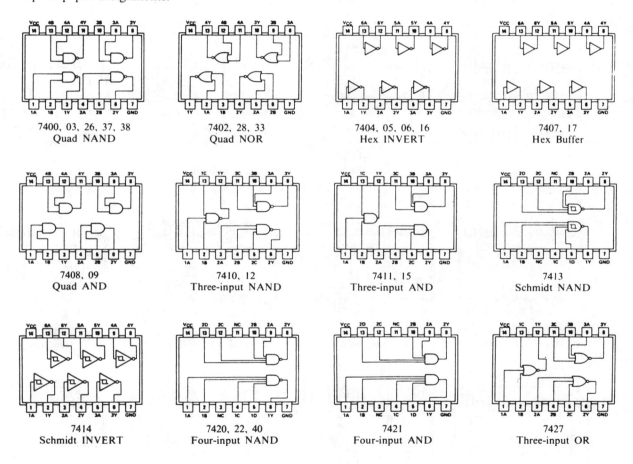

7400, 03, 26, 37, 38
Quad NAND

7402, 28, 33
Quad NOR

7404, 05, 06, 16
Hex INVERT

7407, 17
Hex Buffer

7408, 09
Quad AND

7410, 12
Three-input NAND

7411, 15
Three-input AND

7413
Schmidt NAND

7414
Schmidt INVERT

7420, 22, 40
Four-input NAND

7421
Four-input AND

7427
Three-input OR

306

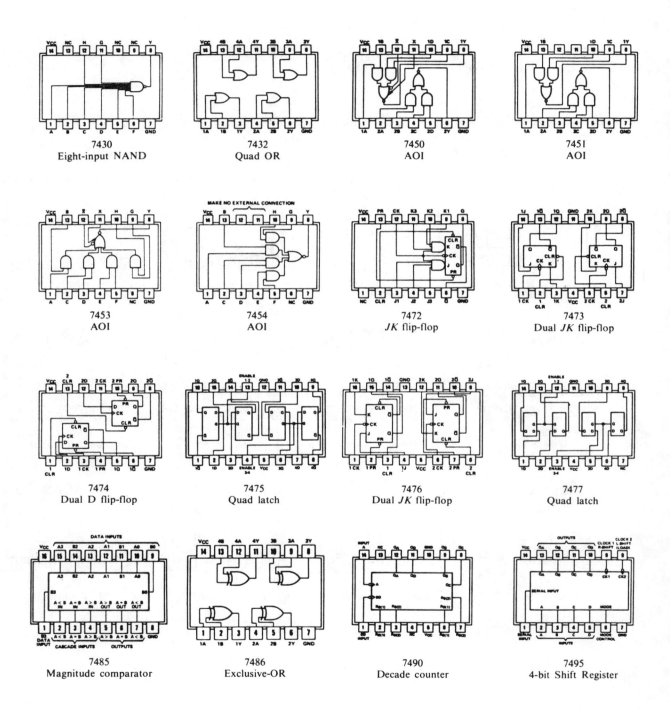

| 7430 | 7432 | 7450 | 7451 |
| Eight-input NAND | Quad OR | AOI | AOI |

| 7453 | 7454 | 7472 | 7473 |
| AOI | AOI | *JK* flip-flop | Dual *JK* flip-flop |

| 7474 | 7475 | 7476 | 7477 |
| Dual D flip-flop | Quad latch | Dual *JK* flip-flop | Quad latch |

| 7485 | 7486 | 7490 | 7495 |
| Magnitude comparator | Exclusive-OR | Decade counter | 4-bit Shift Register |

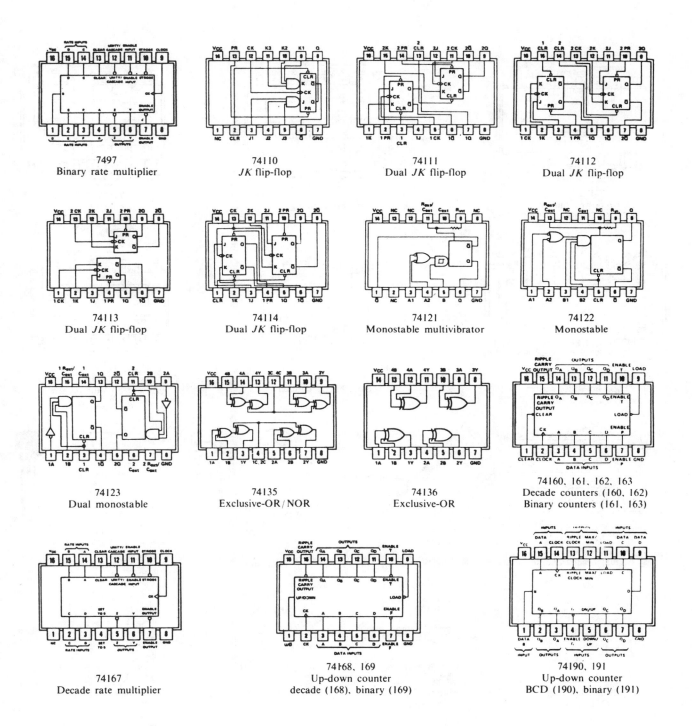

**7497**
Binary rate multiplier

**74110**
*JK* flip-flop

**74111**
Dual *JK* flip-flop

**74112**
Dual *JK* flip-flop

**74113**
Dual *JK* flip-flop

**74114**
Dual *JK* flip-flop

**74121**
Monostable multivibrator

**74122**
Monostable

**74123**
Dual monostable

**74135**
Exclusive-OR/NOR

**74136**
Exclusive-OR

**74160, 161, 162, 163**
Decade counters (160, 162)
Binary counters (161, 163)

**74167**
Decade rate multiplier

**74168, 169**
Up-down counter
decade (168), binary (169)

**74190, 191**
Up-down counter
BCD (190), binary (191)

308

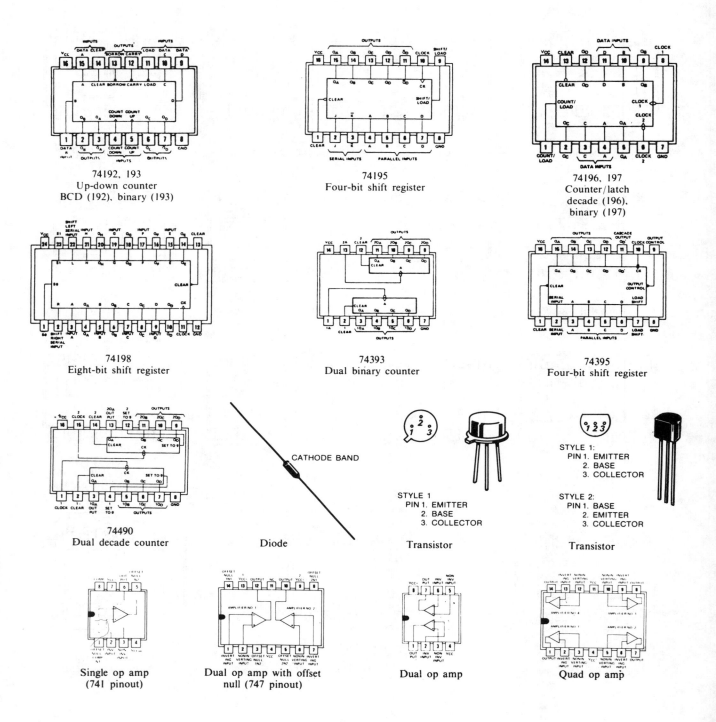

74192, 193
Up-down counter
BCD (192), binary (193)

74195
Four-bit shift register

74196, 197
Counter/latch
decade (196),
binary (197)

74198
Eight-bit shift register

74393
Dual binary counter

74395
Four-bit shift register

74490
Dual decade counter

Diode

CATHODE BAND

Transistor

STYLE 1
PIN 1. EMITTER
2. BASE
3. COLLECTOR

Transistor

STYLE 1:
PIN 1. EMITTER
2. BASE
3. COLLECTOR

STYLE 2:
PIN 1. BASE
2. EMITTER
3. COLLECTOR

Single op amp
(741 pinout)

Dual op amp with offset
null (747 pinout)

Dual op amp

Quad op amp

**NE5018**

NE5018-F,N

## PIN CONFIGURATION

### F,N PACKAGE

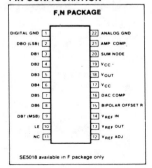

| | | | |
|---|---|---|---|
| DIGITAL GND | 1 | 22 | ANALOG GND |
| DB0 (LSB) | 2 | 21 | AMP COMP |
| DB1 | 3 | 20 | SUM NODE |
| DB2 | 4 | 19 | $V_{CC}$ |
| DB3 | 5 | 18 | $V_{OUT}$ |
| DB4 | 6 | 17 | $V_{CC}$ |
| DB5 | 7 | 16 | DAC COMP |
| DB6 | 8 | 15 | BIPOLAR OFFSET R |
| DB7 (MSB) | 9 | 14 | $V_{REF}$ IN |
| LE | 10 | 13 | $V_{REF}$ OUT |
| NC | 11 | 12 | $V_{REF}$ ADJ |

SE5018 available in F package only

**ADC080X**
**Dual-In-Line Package**

| | | | |
|---|---|---|---|
| $\overline{CS}$ | 1 | 20 | $V_{CC}$ (OR $V_{REF}$) |
| $\overline{RD}$ | 2 | 19 | CLK R |
| $\overline{WR}$ | 3 | 18 | DB0 (LSB) |
| CLK IN | 4 | 17 | DB1 |
| $\overline{INTR}$ | 5 | 16 | DB2 |
| $V_{IN}(+)$ | 6 | 15 | DB3 |
| $V_{IN}(-)$ | 7 | 14 | DB4 |
| A GND | 8 | 13 | DB5 |
| $V_{REF}/2$ | 9 | 12 | DB6 |
| D GND | 10 | 11 | DB7 (MSB) |

**TOP VIEW**

**LF398**

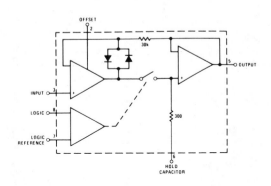

# APPENDIX D. COMPUTER REFERENCE CARDS

This appendix contains computer reference cards. The following pages can be torn out and folded for ready reference.

# Rockwell

## AIM 65 8K BASIC
## Reference Card

## Commands

CLEAR
CONT
FRE
LIST
LOAD
NEW
PEEK
POKE
RUN
SAVE

## Input/Output

DATA
INPUT
PRINT
READ
SPC
TAB

## String Functions

ASC
CHR$
GET
LEFT$
LEN
MID$
RIGHT$
STR$
VAL

## Program Statements

DEF FN.
DIM
END
FOR
GOSUB
GOTO
IF...GOTO
IF...THEN
LET
NEXT
ON...GOSUB
ON...GOTO
REM
RESTORE
RETURN
STOP
USR
WAIT

## Arithmetic Functions

ABS
ATN
COS
EXP
INT
LOG
RND
SGN
SIN
SQR
TAN

## Rockwell Microelectronic Devices
## Sales Offices

WESTERN REGION, U.S.A.
3310 Miraloma Avenue
P.O. Box 3669
Anaheim, CA 92803
Phone: (714) 632-3698

EASTERN REGION, U.S.A.
Carolier Office Building
850-870 U.S. Route 1
North Brunswick, New Jersey 08902
Phone (201) 246-3630

MIDWEST REGION, U.S.A.
1011 E. Touhy—Suite 245
Des Plaines, Illinois 60018
Phone (312) 297-8862

CENTRAL REGION, U.S.A.
Contact Robert O. Whitesell & Associates
6691 East Washington Street
Indianapolis, Indiana 46219
Attn. Milt Gamble, Mgr.
Phone (317) 359-9283

FAR EAST
Rockwell International Overseas Corp
Ichiban-cho Central Building
22-1 Ichiban-cho, Chiyoda-ku
Tokyo 102 Japan
Phone: 265-8808

EUROPE
Rockwell International GmbH
Microelectronic Devices
Fraunhoferstrasse 11
D-8033 Munchen-Martinsried
Germany
Phone: (089) 859-9575

DOC. NO. 29650N55
MAY 1979

## TABLE OF ERROR CODES

| Code | Error |
|------|-------|
| BS | Bad subscript |
| CN | Cannot continue |
| DD | Double dimension |
| FC | Illegal function call |
| ID | Illegal direct |
| LS | String too long |
| NF | NEXT without FOR |
| OD | Out of data |
| OM | Out of memory |
| OV | Overflow |
| RG | RETURN without GOSUB |
| SN | Syntax error |
| ST | String temporaries |
| TM | Type mismatch |
| UF | Undefined function |
| US | Undefined statement |
| /0 | Division by zero |

## MESSAGES

REDO FROM START — A non-decimal character was entered in response to an INPUT function. Re-enter the entire number.

EXTRA IGNORED — An extra parameter was entered in response to an INPUT function.

BREAK — Break command (F1) or STOP command was executed.

## ZERO PAGE PARAMETERS

| Parameter | Hex Address | Decimal Address |
|-----------|-------------|-----------------|
| USR Routine address (L, H) | 04, 05 | 04, 05 |
| Line width | 12 | 18 |
| Input Buffer | 14-50 | 20-88 |
| Pointer to program start (L, H) | 73, 74 | 115, 116 |
| Pointer to variable start (L, H) | 75, 76 | 117, 118 |
| Pointer to array start (L, H) | 77, 78 | 119, 120 |
| Pointer to top of used memory (L, H) | 79, 7A | 121, 122 |
| Pointer to top free space (L, H) | 7B, 7C | 123, 124 |
| Pointer to top memory (L, H) | 7F, 80 | 127, 128 |
| Floating-point accumulator | A9-AE | 169-174 |
| Floating-point argument register | B1-B6 | 177-182 |

## ARITHMETIC FUNCTIONS

| Statement | Syntax/Function | Example |
|-----------|-----------------|---------|
| ABS | ABS (expression) Absolute value of expression | Y = ABS (A + B) |
| ATN | ATN (expression) User-provided function that returns arctangent of the expression (in radians) | PRINT ATN (A) |
| COS | COS (expression) Cosine of the expression (in radians) | A = COS (2, 3) |
| EXP | EXP (expression) Raises the constant e to the power of the variable | B = EXP (C) |
| INT | INT (expression) Evaluates the expression for the largest integer contained | C = INT (X + 3) |
| LOG | LOG (expression) Gives the natural logarithm of the expression | D = LOG (Y − 2) |
| RND | RND (parameter) Generates a random number. Parameters are: <0 seed new sequence =0 return previous random number >0 return new random number | E = RND (1) |
| SGN | SGN (expression) 1 if expression > 0 0 if expression = 0 − 1 if expression < 0 | B = SGN (X + Y) |
| SIN | SIN (expression) Sine of the expression (in radians) | B = SIN (A) |
| SQR | SQR (expression) Square root of the expression | C = SQR (D) |
| TAN | TAN (expression) Tangent of the expression (in radians) | D = TAN (3.14) |

## BASIC INSTRUCTION FORMAT

00A₂A₂A₂A₀N₃N₁N₀XX...XX00A₂A₂A₂A₀...

Where:
00 = Line delimiter; i.e., start of new statement line

A₁A₂A₂A₀ = Address of the start of the next statement, in binary

N₃N₁N₀ = Statement line number, in binary

XX....XX = Tokens (statement codes) and data, in ASCII and BASIC encoded symbols

## OPERATORS

| Symbol | Function |
|---|---|
| = | Assignment, or equality test |
| − | Negation or subtraction |
| + | Addition or string concatenation |
| * | Multiplication |
| / | Division (floating point result) |
| ↑ (F3) | Exponentiation |
| NOT | One's complement (integer) |
| AND | Bitwise AND (integer) |
| OR | Bitwise OR (integer) |
| XOR | Bitwise exclusive OR (integer) |
| = | Equal |
| < | Less than |
| > | Greater than |
| <= | Less than or equal |
| >= | Greater than or equal |
| <> | Not equal |

The precedence of operators is:
(1) Expressions in parentheses
(2) Exponentiation (A↑B)
(3) Negation (−X)
(4) *, /
(5) +, −
(6) Relational operators [=,<>,<,>,<=,>=]
(7) NOT
(8) AND
(9) OR
(10) XOR

## SPECIAL CHARACTERS

| | |
|---|---|
| RETURN | Ends every line typed in |
| LF | Advances printer paper one line |
| DEL | Erases last character typed |
| :(colon) | Separates statements typed on the same line |
| ? | Equivalent to PRINT |
| F1 | Break in execution |
| ESC | Returns control to monitor |
| _ | Erases current line |
| @,` | Causes print even if printer control is turned off |
| $ | Suffix, specifies string variable |
| % | Suffix, specifies integer variable |

## COMMANDS

| Statement | Syntax/Function | Example |
|---|---|---|
| CLEAR | Clear program variables | CLEAR |
| CONT | Continue program execution | CONT |
| FRE | FRE (expression) Gives memory free space not used by BASIC | PRINT FRE (0) |
| LIST | LIST [[start line]-[end line]]] List program lines at terminal | LIST 100-1000 |
| LOAD | LOAD input device filename tape no. Load a program file. | LOAD IN = T F = NAME1 T = 1 |
| NEW | Delete current program and variables. | NEW |
| PEEK | PEEK (address) Reads a byte in decimal from memory at specified decimal address | PRINT PEEK (2000) |
| POKE | POKE address, byte Puts byte specified in decimal into decimal memory location specified | POKE 23100, 255 |
| RUN | RUN (line number) Run a program (from line number) | RUN / RUN 50 |
| SAVE | SAVE output device filename, tape no. Save the program in memory with name "filename." | SAVE / OUT = T F = PROG T = 1 |

## PROGRAM STATEMENTS

| Statement | Syntax/Function | Example |
|---|---|---|
| DEF FN | DEF FNx [(argument list)] = expression Define an arithmetic or string function | DEF FNA (X,Y) = SQR (X*X + Y*Y) |
| DIM | DIM variable (size 1, [ size 2 ...]). Allocate space for arrays. | DIM A (3), B$ (10, 2, 3) |
| END | END Stop program and return to BASIC command level. | END |
| FOR | FOR variable = expression TO expression [STEP expression] Used with NEXT statement to repeat a sequence of program lines. The variable is incremented by the value of STEP. | FOR I = 1 TO 5 STEP .5 .... |
| GOSUB | GOSUB line number Call a BASIC subroutine by branching to the specified line number. See RETURN. | GOSUB 210 |
| GOTO | GOTO line number Branch to specified line number | GOTO 90 |
| IF...GOTO | IF expression GOTO line number The relation X<Y is tested; if true, the GOTO clause is executed. If false the GOTO clause is not executed. | IF X <Y GOTO 100 |
| IF...THEN | IF expression THEN statement [:statement]... The relation X<Y is tested. If true, the THEN clause is executed. If false, the THEN clause is not executed. | IF X<Y THEN Y = X |
| LET | [LET] variable = expression Assign a value to a variable | LET X = I + 5 |
| NEXT | NEXT variable [,variable].... Delimits the end of a FOR loop | NEXT |
| ON...GOSUB | ON expression GOSUB line [,line]... GOSUBs to statement specified by expression. (If J + 1 = 1, to 20; if J + 1 = 2, to 20 if J + 1 = 3, to 40) | ON J + 1 GOSUB 20,20,40 |
| ON...GOTO | ON expression GOTO line [,line]... Branches to statement specified by I (To 20 if I = 1; to 30 if I = 2) | ON I GOTO 20,30 |
| REM | REM any text Allows user to insert comments in program (not executed). Note:";" does not terminate a REM statement. | REM comment |
| RESTORE | RESTORE Resets DATA pointer so that DATA statements may be re-read | RESTORE |
| RETURN | RETURN Return from subroutine to statement following last GOSUB performed | RETURN |
| STOP | STOP Stop program execution, print BREAK message, and return to COMMAND mode. | STOP |
| USR | USRn (argument) Calls the user's machine language subroutine with the specified argument. | PRINT USR (27000) |
| WAIT | WAIT address, mask [,select] Waits for input on a hardware port | WAIT 21, 1 |

## INPUT OUTPUT STATEMENTS

| Statement | Syntax/Function | Example |
|---|---|---|
| DATA | DATA item [,item ...] Specifies data to be used in a READ statement | DATA 2,3, "PLUS", 4 |
| INPUT | INPUT [ !] ["prompt string literal";] variable [, variable]... Read data from the keyboard If ! is included, the input data is printed regardless of printer control state | INPUT "VALUES"; A, B |
| PRINT | PRINT [ !] expression [, expression] Display/print data at the terminal. If ! is included, the data is printed regardless of the printer control state | PRINT X;Y / A$, B$ / PRINT X, Y / PRINT "VALUE"; X |
| READ | READ variable [, variable] Reads data into specified variables from DATA statement | READ I, X, A$ |
| SPC | SPC (expression) Used in PRINT statement to print spaces | PRINT SPC (5) |
| TAB | TAB (expression) Used in PRINT statement to tab start of print to specified position. | PRINT TAB (20) |

## STRING FUNCTIONS

| Statement | Syntax/Function | Example |
|---|---|---|
| ASC | ASC (string expression) Returns the ASCII value of the first character of a string | PRINT ASC (A$) |
| CHR$ | CHR$ (expression) Returns one character, the ASCII equivalent of the expression | PRINT CHR$ (48) |
| GET | GET string variable Inputs a single character from the keyboard. Continues execution if no character is available. | GET A$ |
| LEFT$ | LEFT$ (string expression, length) Returns leftmost length characters of the string expression | B$ = LEFT$ (X$, 8) |
| LEN | LEN (string expression) Returns the length of a string | PRINT LEN (B$) |
| MID$ | MID$ (string expression, start [,length]) Returns characters from the middle of the string starting at the position specified to the end of the string or for length characters MID$ may also be used to assign a substring inside a string | A$ = MID$ (X$, 5, 10) |
| RIGHT$ | RIGHT$ (string expression, length) Returns rightmost length characters of the string expression | C$ = RIGHT$ (X$, 8) |
| STR$ | STR$ (expression) Converts a numeric expression to a string | PRINT STR$ (35) |
| VAL | VAL (string expression) Converts the string representation of a number to its numeric value | PRINT VAL ("3.1") |

# Rockwell AIM 65 Summary Card

## R6522 PERIPHERAL CONTROL REGISTER (PCR), LOC. $A00C

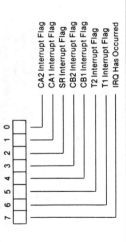

7 6 5 4 3 2 1 0

- CA1 Control
- CA2 Control
- CB1 Control
- CB2 Control

**CA1 CONTROL**

PCR0 = 0 The CA1 Interrupt Flag (IFR1) will be set by a negative transition (high to low) on the CA1 pin.

= 1 The CA1 Interrupt Flag (IFR1) will be set by a positive transition (low to high) on the CA1 pin.

**CA2 CONTROL**

| PCR3 | PCR2 | PCR1 | Mode |
|---|---|---|---|
| 0 | 0 | 0 | CA2 negative edge interrupt (IFR0/ORA clear) mode — Set CA2 interrupt flag (IFR0) on a negative transition of the CA2 input signal. Clear IFR0 on a read or write of the ORA or by writing logic 1 into IFR0. |
| 0 | 0 | 1 | CA2 negative edge interrupt (IFR0 clear) mode — Set IFR0 on a negative transition of the CA2 input signal. Clear IFR0 by writing logic 1 into IFR0. |
| 0 | 1 | 0 | CA2 positive edge interrupt (IFR0/ORA clear) mode — Set CA2 interrupt flag (IFR0) on a positive transition of the CA2 input signal. Clear IFR0 on a read or write of the ORA or by writing logic 1 into IFR0. |
| 0 | 1 | 1 | CA2 positive edge interrupt (IFR0 clear) mode — Set IFR0 on a positive transition of the CA2 input signal. Clear IFR0 by writing logic 1 into IFR0. |
| 1 | 0 | 0 | CA2 handshake output mode — Set CA2 output low on a read or write of the Peripheral A Output Register. Reset CA2 high with an active transition on CA1. |
| 1 | 0 | 1 | CA2 pulse output mode — CA2 goes low for one cycle following a read or write of the Peripheral A Output Register. |
| 1 | 1 | 0 | CA2 low output mode — The CA2 output is held low in this mode. |
| 1 | 1 | 1 | CA2 high output mode — The CA2 output is held high in this mode. |

**CB1 CONTROL**

PCR4 = 0 The CB1 Interrupt Flag (IFR4) will be set by a negative transition (high to low) on the CB1 pin.

= 1 The CB1 Interrupt Flag (IFR4) will be set by a positive transition (low to high) on the CB1 pin.

**CB2 CONTROL**

| PCR7 | PCR6 | PCR5 | Mode |
|---|---|---|---|
| 0 | 0 | 0 | CB2 negative edge interrupt (IFR3/ORB clear) mode — Set CB2 interrupt flag (IFR3) on a negative transition of the CB2 input signal. Clear IFR3 on a read or write of the ORB or by writing logic 1 into IFR3. |
| 0 | 0 | 1 | CB2 negative edge interrupt (IFR3 clear) mode — Set IFR3 on a negative transition of the CB2 input signal. Clear IFR3 by writing logic 1 into IFR3. |
| 0 | 1 | 0 | CB2 positive edge interrupt (IFR3/ORB clear) mode — Set CB2 interrupt flag (IFR3) on a positive transition of the CB2 input signal. Clear IFR3 on a read or write of the ORB or by writing logic 1 into IFR3. |
| 0 | 1 | 1 | CB2 positive edge interrupt (IFR3 clear) mode — Set IFR3 on a positive transition of the CB2 input signal. Clear IFR3 by writing logic 1 into IFR3. |
| 1 | 0 | 0 | CB2 handshake output mode — Set CB2 output low on a write of the Peripheral B Output Register. Reset CB2 high with an active transition on CB1. |
| 1 | 0 | 1 | CB2 pulse output mode — CB2 goes low for one cycle following a read or write of the Peripheral B Output Register. |
| 1 | 1 | 0 | CB2 low output mode — The CB2 output is held low in this mode. |
| 1 | 1 | 1 | CB2 high output mode — The CB2 output is held high in this mode. |

## R6522 INTERRUPT FLAG REGISTER (IFR), LOC. $A00D

7 6 5 4 3 2 1 0

- CA2 Interrupt Flag
- CA1 Interrupt Flag
- SR Interrupt Flag
- CB2 Interrupt Flag
- CB1 Interrupt Flag
- T2 Interrupt Flag
- T1 Interrupt Flag
- IRQ Has Occurred

| IFR Bit | Set By | Cleared By |
|---|---|---|
| 0 | Active transition on CA2 | Reading or writing the ORA ($A001 or $A00F) |
| 1 | Active transition on CA1 | Reading or writing the ORA ($A001 or $A00F) |
| 2 | Completion of eight shifts | Reading or writing the SR ($A00A) |
| 3 | Active transition on CB2 | Reading or writing the ORB ($A000) |
| 4 | Active transition on CB1 | Reading or writing the ORB ($A000) |
| 5 | Time-out of Timer 2 | Reading T2C-L ($A008) or writing T2C-H ($A009) |
| 6 | Time-out of Timer 1 | Reading T1C-L ($A004) or writing T1L-H ($A005 or $A007) |
| 7 | Any IFR bit set with its corresponding IER bit also set | Clearing IFR0-IFR6 ($A00D) or IER0-IER6 ($A00E) |

## R6522 INTERRUPT ENABLE REGISTER (IER), LOC. $A00E

7 6 5 4 3 2 1 0

- CA2 Interrupt Enable
- CA1 Interrupt Enable
- SR Interrupt Enable
- CB2 Interrupt Enable
- CB1 Interrupt Enable
- T2 Interrupt Enable
- T1 Interrupt Enable
- IER Set/Clear Control

**INTERRUPT ENABLE BITS (IER0-6)**

IERn = 0 Disable interrupt

= 1 Enable interrupt

**IER SET/CLEAR CONTROL (IER7)**

IER7 = 0 For each data bus bit set to logic 1, clear corresponding IER bit.

= 1 For each data bus bit set to logic 1, set corresponding IER bit.

Note: IER7 is active only when $R/\overline{W} = L$; when $R/\overline{W} = H$, IER7 will read logic 1.

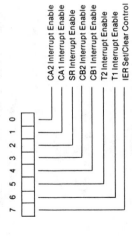

# Rockwell Microelectronic Devices Sales Offices

**SOUTHWEST REGION, U.S.A.**
3310 Miraloma Avenue
P.O. Box 3669
Anaheim, CA 92803
Phone (714) 632-3698

**NORTHWEST REGION, U.S.A.**
Rockwell International
1601 Civic Center Drive, Suite 203
Santa Clara, CA 95050
Phone (408) 984-6070

**EASTERN REGION, U.S.A.**
Caroller Office Building
850-870 U.S. Route 1
North Brunswick, New Jersey 08902
Phone (201) 246-3630

**MIDWEST REGION, U.S.A.**
1011 E. Touhy — Suite 245
Des Plaines, Illinois 60018
Phone (312) 297-8862

**CENTRAL REGION, U.S.A.**
Contact Robert O. Whitesell & Associates
6691 East Washington Street
Indianapolis, Indiana 46219
Attn. Milt Gamble, Mgr.
Phone (317) 359-9283

**FAR EAST**
Rockwell International Overseas Corp
Itohpia Hirakawa-cho Bldg.
7-6 Hirakawa-cho 2-chome
Chiyoda-ku, Japan
Phone (03) 265-8806

**EUROPE**
Rockwell International GmbH
Microelectronic Devices
Fraunhoferstrasse 11
D-8033 Munchen-Martinsried
Germany
Phone (089) 859 9575

DOC. NO. 29650N51
REV. 2, JANUARY 1980

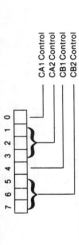

# AIM 65 MONITOR COMMANDS

## MAJOR FUNCTION ENTRY COMMANDS

[RESET] — Enter and Initialize Monitor
ROCKWELL AIM 65

E — Enter and Initialize Editor
<E>

T — Re-enter Text Editor at Top of Text
<T>
TOP LINE OF TEXT

N — Enter Assembler
<N>

5 — Enter and Initialize BASIC Interpreter
<5>

6 — Re-enter BASIC Interpreter
<6>

## INSTRUCTION ENTRY AND DISASSEMBLY COMMANDS

I — Enter Mnemonic Instruction Entry Mode
<I>
AAAA[*] = [ADDRESS]
AAAA XX [OPCODE][HEX OPERAND]
AAAA XX XX XX

K — Disassemble Memory
<K> * = [ADDRESS]
/[DECIMAL NUMBER]
AAAA XX OPCODE HEX OPERAND

## DISPLAY/ALTER REGISTER COMMANDS

* — Alter Program Counter
<*> = [ADDRESS]

A — Alter Accumulator
<A> = [BYTE]

X — Alter X Register
<X> = [BYTE]

Y — Alter Y Register
<Y> = [BYTE]

P — Alter Processor Status
<P> = [BYTE]

S — Alter Stack Pointer
<S> = [BYTE]

R — Display Register Values
<R>
**** PS AA XX YY SS
0200 00 00 01 02 FF

## DISPLAY/ALTER MEMORY CONTENTS

M — Display Specified Memory Locations
<M> = [ADDRESS]XX XX XX XX

SPACE — Display Next 4 Memory Locations
< >AAAA XX XX XX XX

/ — Alter Current Memory Locations
</>AAAA XX XX XX XX

## LOAD/DUMP MEMORY COMMANDS

L — Load Object Code into Memory
<L> IN = [INPUT DEVICE]

D — Dump Memory
<D>
FROM = [ADDRESS]TO = [OUTPUT DEVICE]
OUT = [OUTPUT DEVICE]
MORE?[Y, N]

## BREAKPOINT MANIPULATION COMMANDS

# — Clear All Breakpoints
<#>OFF

4 — Toggle Breakpoint Enable
<4>OFF/ON

B — Set/Clear Breakpoint Address
<B> BRK/[0, 1, 2, 3] = [ADDRESS]

? — Display Breakpoint Addresses
<?>
AAAA AAAA AAAA AAAA

# AIM 65 MONITOR COMMANDS (Continued)

## EXECUTION/TRACE CONTROL COMMANDS

G — Start Execution of User's Program
<G>/[DECIMAL NUMBER]

Z — Toggle Instruction Trace Mode
<Z>ON/OFF

V — Toggle Register Trace Mode
<V>ON/OFF

H — Trace Program Counter History
<H>
AAAA
<N>
AAAA
.
.
.
AAAA

## CONTROL PERIPHERAL DEVICES

CTRL PRINT — Toggle Printer On/Off
<CTRL><PRINT>

PRINT — Print Display Contents
<PRINT>

LF — Advance Printer Paper
<LF>

1 — Toggle Tape 1 Control On/Off
<1>

2 — Toggle Tape 2 Control On/Off
<2>

3 — Tape Verify Block Checksum
<3>IN = [T]F = [FILE NAME]T = [1,2]

## USER FUNCTION COMMANDS

F1 — Call User Function 1 (through loc. $010C)
<F1>

F2 — Call User Function 2 (through loc. $010F)
<F2>

F3 — Call User Function 3 (through loc. $0112)
<F3>

## AIM 65 COMMAND DEFINITIONS

[ADDRESS] Hexadecimal address, one to four characters

[BYTE] Two-digit hexadecimal value from 00 to FF.

[DECIMAL NUMBER] A two-digit decimal number in the range 00 to 99.

[FILE NAME] A string of 1 to 5 characters.

[HEX OPERAND] The instruction operand.

| Addressing Mode | Operand Format |
| --- | --- |
| Accumulator | A |
| Immediate | #HH |
| Zero Page | HH |
| Zero Page, X | HH, X or HHX |
| Zero Page, Y | HH, Y or HHY |
| Absolute | HHHH |
| Absolute, X | HHHH, X or HHHHX |
| Absolute, Y | HHHH, Y or HHHHY |
| Relative | HH or HHHH |
| (Indirect, X) | (HH,X) or (HHX) or (HH,X or (HHX |
| (Indirect), Y | (HH),Y or (HH)Y |
| (Absolute Indirect) | (HHHH) |

[INPUT DEVICE] RETURN or SPACE — AIM 65 Keyboard (S2 = KB) or TTY Keyboard (S2 = TTY)
M — Memory
T — Audio Tape, AIM 65 format
K — Audio Tape, KIM-1 format
L — TTY Paper Tape Reader
U — User-defined input device

[MNEMONIC OPCODE] A three-letter mnemonic abbreviation.

[OUTPUT DEVICE] RETURN or SPACE — AIM 65 Display/Printer (S2 = KB) or TTY Printer (S2 = TTY)
P — AIM 65 Printer
X — Dummy
T — Audio Tape, AIM 65 format
K — Audio Tape, KIM-1 format
L — TTY Paper Tape Punch
U — User-defined output device

# AIM 65 TEXT EDITOR COMMANDS

## ENTER AND EXIT EDITOR COMMANDS

E — Enter and Initialize Editor
<E>
EDITOR
FROM = [ADDRESS] TO = [ADDRESS]
IN = [INPUT DEVICE]
Note: Defaults are TO = $0200,
FROM = Last contiguous RAM, IN = Keyboard

Q — Exit the Text Editor and Return to Monitor
= <Q>

## LINE ORIENTED COMMANDS

R — Read Lines into Text Buffer from Input Device
= <R>
IN = [INPUT DEVICE]

I — Insert One Line of Text Ahead of Active Line
= <I>
INSERTED TEXT LINE
ACTIVE LINE OF TEXT

K — Delete Current Line of Text
= <K>
DELETED LINE OF TEXT
ACTIVE LINE OF TEXT

U — Move the Text Pointer Up One Line
= <U>
PRIOR LINE OF TEXT

D — Move the Text Pointer Down One Line
= <D>
NEXT LINE OF TEXT

T — Move the Text Pointer to the Top of the Text
= <T>
TOP LINE OF TEXT

B — Move the Text Pointer to the Bottom of the Text
= <B>
BOTTOM LINE OF TEXT

L — List Lines of Text to Output Device
= <L>
/[DECIMAL NUMBER]

SPACE — Display the Active Line
= < >
ACTIVE LINE OF TEXT

## STRING ORIENTED COMMANDS

F — Find a Character String
= <F>
[CHARACTER STRING]
LINE CONTAINING CHARACTER STRING

C — Change a Character String
= <C>
[OLD STRING]
TO = [NEW STRING]
LINE CONTAINING OLD STRING
SAME LINE, WITH NEW STRING

## ASSEMBLER ERROR CODES

01 Undefined Symbol or Label
02 Label Previously Used or Forward Reference to Page 0 Symbol
03 Illegal or Missing Opcode
04 Invalid Address
05 Accumulator Mode Invalid
06 Forward Reference to Page Zero Symbol
07 Invalid Source Statement
08 Label Begins With Numeric Character
09 Label Longer Than Six Characters
10 Non-Alphanumeric Label or Opcode
11 Forward Reference in Equate
12 Invalid Index — Must Be X or Y
13 Invalid Expression
14 Undefined Assembler Directive
15 Invalid Page 0 Operand
17 Relative Branch Out of Range
18 Illegal Operand Type for This Instruction
19 Invalid Indirect Pointer
20 A, X, Y, S and P are Reserved Names
21 Location Counter Negative — Reset P to 0

# AIM 65 ASSEMBLER

## ASSEMBLER COMMAND SUMMARY

```
<N>
ASSEMBLER
FROM = [ADDRESS] TO = [ADDRESS]
IN = [INPUT DEVICE]
LIST? [Y, N]
LIST-OUT = [OUTPUT DEVICE]
OBJ? [Y, N]    Note: N = Object code to Memory
OBJ-OUT = [OUTPUT DEVICE]    Note: Prompts only on Y response to OBJ?
PASS 1
SYM TBL OVERFLOW  }  Displayed only if Symbol Table overflows
ASSEMBLER
PASS 2
= = AAAA LABEL              Displayed only if
OBJECT CODE  MNEMONIC OPCODE   LIST?Y, or LIST?N
SYMBOLIC OPERAND  ;COMMENT     and error detected
**ERROR NN  Note: Error code displayed only on error
ERRORS = MMMM  Decimal count of errors detected
```

## ASSEMBLER EXPRESSIONS

### ELEMENTS

Numeric constants — may be written in one of four bases.

| Prefix Character | Base |
|---|---|
| (none) | 10 (Decimal) |
| $ | 16 (Hexadecimal) |
| @ | 8 (Octal) |
| % | 2 (Binary) |

### OPERATORS

| Type | Operator | Operation |
|---|---|---|
| Arithmetic | + | Addition |
| Arithmetic | - | Subtraction |
| Special | > | High-Byte Selection |
| Special | < | Low-Byte Selection |

Operators < and > truncate a two-byte value to its low or high byte, respectively.

## ASSEMBLER DIRECTIVES

**=** — Assigns the value of an operand containing no forward references to either a symbol or the location counter.
{SYMBOL / *} = Operand

**.BYTE** — Assigns multiple ASCII strings or expressions to consecutive single byte memory locations.
.BYT Expression, Expression,... Expression

**.WORD** — Assigns multiple expression operands to consecutive memory locations in low-byte, high-byte order.
.WOR Expression, Expression, ... Expression

**.DBYTE** — Assigns multiple expression operands to consecutive double byte (16 bits) memory locations in high-byte, low-byte order.
.DBY Expression, Expression, ... Expression

**.PAGE** — Generates a title under a dashed line.
.PAG {'NEW TITLE' / BLANK}   (New Title)(No Change of Title)(Blanks Title)

**.SKIP** — Generates one blank line.
.SKI

**.OPT** — Controls assembly listings. All are optional and can be specified in any order or in separate statements.
.OPT {LIS/NOL}, {GEN/NOG}, {ERR/NOE}

**.FILE** — Last record in a multiple file source program (except the last file) which points to the continuation file.
.FIL File Name

**.END** — Last record in a single or multiple source file.
.END

# AIM 65 SUBROUTINE SUMMARY

| Sub. Name | Entry Addr. | Registers Altered | Function |
|---|---|---|---|
| BLANK | E83E | A | Outputs one SP to D/P |
| BLANK2 | E83B | A | Outputs two SP's to D/P |
| CLR | EB44 | A | Clears D/P pointers |
| CRCK | EA24 | A | Outputs print buffer to Printer |
| CRLF | E9F0 | A | Outputs CR, LF & NUL to AOD |
| CRLOW | EA13 | A | Outputs CR & LF to D/P |
| CUREAD | FE83 | A | Inputs one ASCII character from KB to A, displays cursor |
| DEBK1 | ED2C | A | Generates a five-millisecond delay |
| DUMPTA | E56F | A | Opens Audio Tape output file |
| EQUAL | E7D8 | A | Outputs "=" to D/P |
| FROM | E7A3 | A,X,Y | Outputs "FROM =" to D/P and enters address |
| GETTAP | EE29 | A,Y | Inputs one character from Audio Tape |
| HEX | EA7D | A | Converts a hex number in A from ASCII to binary, and puts result in the LSD of A, with zero in MSD of A. |
| INALL | E993 | A | Inputs one ASCII character from AID to A |
| INLOW | E8F8 | A | Selects KB input in INFLG |
| LL | E8FE | A | Selects input from KB and output to D/P |
| LOADTA | E32F | A | Searches for audio input file |
| NOUT | EA51 | A | Converts a hex number in LSD of A from binary to ASCII, and outputs it to AOD |
| NUMA | EA46 | A | Converts two hex numbers in A from binary to ASCII, and outputs them to AOD, MSD first. |
| OUTALL | E9BC | A | Outputs ASCII character in A to AOD |
| OUTLOW | E901 | A | Selects output to D/P |
| OUTPUT | E97A | A | Outputs ASCII character in A to D/P |
| PACK | EA84 | A | Converts a hex number in A from ASCII to binary, and puts result in the LSD of A, with the result of the last call to PACK or HEX in the MSD of A. |
| PHXY | EB9E | | Push X and Y Registers onto Stack |
| PLXY | EBAC | X,Y | Pull X and Y Registers from Stack |
| PSL1 | E837 | A | Outputs "/" to D/P |
| QM | E7D4 | A | Outputs "?" to D/P |
| RBYTE | E3FD | A | Inputs two ASCII characters from AID; if hex, converts to binary with result in A |
| RCHEK | E907 | A,X,Y | Scans KB, returns to Monitor on ESC, to caller on no entry, wait on SP |
| RDRUB | E95F | A,Y | Inputs one ASCII character from KB to A, with echo to D/P. Allows DEL, if Y ≠ 0. |
| READ | E93C | A | Inputs one ASCII character from KB to A |
| RED1 | FE96 | A | Inputs one character from KB to A, with echo to D/P |
| REDOUT | E973 | A | Inputs one ASCII character from KB to A, with echo to D/P, displays cursor |
| SEMI | E9BA | A | Outputs ";" to AOD |
| TAISET | EDEA | | Sets up Audio Tape input, detects five SYN characters |
| TAOSET | F21D | | Sets up Audio Tape output, issues (GAP) x 4 SYN characters |
| TIBY1 | ED53 | | Loads a block of 80 bytes from Audio,Tape |
| TO | E7A7 | A,X,Y | Outputs "TO" to D/P and enters address |
| WHEREI | E848 | A,X,Y | Sets up the AID and loads INFLG |
| WHEREO | E871 | A,X,Y | Sets up the AOD and loads OUTFLG |

**ABBREVIATIONS**

D/P = Display/Printer
AOD = Active Output Device
AID = Active Input Device

## USER R6522 VERSATILE INTERFACE ADAPTER (VIA)

### R6522 MEMORY ASSIGNMENTS

| Location | Function | | |
|---|---|---|---|
| A000 | Port B Output Data Register (ORB) | | |
| A001 | Port A Output Data Register (ORA) | } Controls handshake | |
| A002 | Port B Data Direction Register (DDRB) | } 0 = Input | |
| A003 | Port A Data Direction Register (DDRA) | } 1 = Output | |

| | Timer | R/W̄ = L | R/W̄ = H |
|---|---|---|---|
| A004 | T1 | Write T1L-L<br>Clear T1 Interrupt Flag | Read T1C-L<br>Read T1C-H |
| A005 | T1 | Write T1L-H & T1C-H<br>T1L-L → T1C-L<br>Clear T1 Interrupt Flag | Read T1C-H |
| A006 | T1 | Write T1L-L | Read T1L-L |
| A007 | T1 | Write T1L-H<br>Clear T1 Interrupt Flag | Read T1L-H |
| A008 | T2 | Write T2L-L | Read T2C-L<br>Clear T2 Interrupt Flag |
| A009 | T2 | Write T2C-H<br>T2L-L → T2C-L<br>Clear T2 Interrupt Flag | Read T2C-H |
| A00A | Shift Register (SR) | | |
| A00B | Auxiliary Control Register (ACR) | | |
| A00C | Peripheral Control Register (PCR) | | |
| A00D | Interrupt Flag Register (IFR) | | |
| A00E | Interrupt Enable Register (IER) | | |
| A00F | Port A Output Data Register (ORA) | No effect on handshake | |

### R6522 AUXILIARY CONTROL REGISTER (ACR), LOC. $A00B

```
7 6 5 4 3 2 1 0
          | | └─ Port A Latch Enable
          | └─── Port B Latch Enable
        └─────── Shift Register Control
      └───────── Timer 2 Control
  └───────────── Timer 1 Control
```

**PORT A LATCH ENABLE**

$ACR0 = 1$ Port A latch is enabled to latch input data when CA1 Interrupt Flag (IFR1) is set.

$= 0$ Port A latch is disabled, reflects current data on PA pins.

**PORT B LATCH ENABLE**

$ACR1 = 1$ Port B latch is enabled to latch the voltage on the pins for the input lines or the ORB contents for the output lines when CB1 Interrupt Flag (IFR4) is set.

$= 0$ Port B latch is disabled, reflects current data on PB pins.

**SHIFT REGISTER CONTROL**

| ACR4 | ACR3 | ACR2 | Mode |
|---|---|---|---|
| 0 | 0 | 0 | Shift Register Disabled. |
| 0 | 0 | 1 | Shift in under control of Timer 2. |
| 0 | 1 | 0 | Shift in under control of ø2. |
| 0 | 1 | 1 | Shift in under control of external clock. |
| 1 | 0 | 0 | Free-running output at rate determined by Timer 2. |
| 1 | 0 | 1 | Shift out under control of Timer 2. |
| 1 | 1 | 0 | Shift out under control of ø2. |
| 1 | 1 | 1 | Shift out under control of external clock. |

**TIMER 2 CONTROL**

$ACR5 = 0$ T2 acts as an interval timer in the one-shot mode.

$= 1$ T2 counts a predetermined number of pulses on PB6.

**TIMER 1 CONTROL**

| ACR7 | ACR6 | Mode |
|---|---|---|
| 0 | 0 | T1 one-shot mode — Generate a single time-out interrupt each time T1 is loaded. Output to PB7 disabled. |
| 0 | 1 | T1 free-running mode — Generate continuous interrupts. Output to PB7 disabled. |
| 1 | 0 | T1 one-shot mode — Generate a single time-out interrupt and an output pulse on PB7 each time T1 is loaded. |
| 1 | 1 | T1 free-running mode — Generate continuous interrupts and a square wave output on PB7. |

## AIM 65 MEMORY MAP

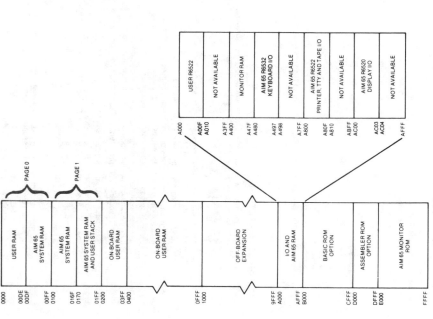

|  |  |
|---|---|
| 0000 | USER RAM (PAGE 0) |
| 00DE/00DF | |
| 00FF | AIM 65 SYSTEM RAM |
| 0100 | (PAGE 1) |
| 016F/0170 | AIM 65 SYSTEM RAM |
| 01FF | AIM 65 SYSTEM RAM AND USER STACK |
| 0200 | ON BOARD USER RAM |
| 03FF | |
| 0400 | ON BOARD USER RAM |
| 0FFF | |
| 1000 | OFF BOARD EXPANSION |
| 9FFF | |
| A000 | I/O AND AIM 65 RAM |
| AFFF | |
| B000 | BASIC ROM OPTION |
| CFFF | |
| D000 | ASSEMBLER ROM OPTION |
| DFFF | |
| E000 | AIM 65 MONITOR ROM |
| FFFF | |

I/O and AIM 65 RAM expansion (A000–AFFF):

|  |  |
|---|---|
| A000 | USER R6522 |
| A00F | |
| A010 | NOT AVAILABLE |
| A3FF | |
| A400 | MONITOR RAM |
| A47F | |
| A480 | AIM 65 R6532 KEYBOARD I/O |
| A497 | |
| A498 | NOT AVAILABLE |
| A7FF | |
| A800 | AIM 65 R6522 PRINTER, TTY AND TAPE I/O |
| A80F | |
| A810 | NOT AVAILABLE |
| ABFF | |
| AC00 | AIM 65 R6520 DISPLAY I/O |
| AC03 | |
| AC04 | NOT AVAILABLE |
| AFFF | |

## AIM 65 USER-ALTERABLE ADDRESSES

| Location | Name | Bytes | Description |
|---|---|---|---|
| 0108 | UIN | 2 | Vector to User Input Handler |
| 010A | UOUT | 2 | Vector to User Output Handler |
| 010C | KEYF1 | 3 | JMP to User Function 1 |
| 010F | KEYF2 | 3 | JMP to User Function 2 |
| 0112 | KEYF3 | 3 | JMP to User Function 3 |
| A400 | IRQV4 | 2 | Vector to IRQ after Monitor Interrupt Routine |
| A402 | NMIV2 | 2 | Vector to NMI Interrupt Routine |
| A404 | IRQV2 | 2 | Vector to IRQ Interrupt Routine |
| A406 | DILINK | 2 | Vector to Display Routine |
| A408 | TSPEED | 1 | Audio Tape Speed<br>Default = $C7 (AIM 65)<br>Options = $5A (KIM-1 x 1)<br>= $5B (KIM-1 x 3) |
| A409 | GAP | 1 | Audio Tape Gap<br>Default = $08 = 32 SYN characters<br>Option = $80 for Assembler input & Editor update |

## APPENDIX E.  PROGRAM LISTINGS

```
        'PLT14 7/23/81'
;WRITTEN BY J. AVERY
;
;THIS PROGRAM PLOTS
;THE VALUES IN THE
;FIRST ARRAY USED IN
;THE BASIC ARRAY
;SPACE.  THE ARRAY
;MUST BE INTEGER,
;AND SCALED BETWEEN
;O AND 139.  THE OTH
;ELEMENT MUST HOLD
;THE # OF POINTS TO
;BE PLOTTED.
;**** THE ONLY WAY
;TO STOP THE PROGRAM
;IS WITH THE RESET
;BUTTON!!!
;MUST RESERVE SPACE
;IN UPPER MEMORY IN
;BASIC FOR PROGRAM
;CURRENTLY, 3580
;WORKS WELL
;
;FOR MORE INFO, SEE
;MONITOR PRINTER
;ROUTINES.
;
;PAGE O LOCATION:
ARRAY=$FE
.PAGE
*=$EDO
;PROGRAM ENTRY
;
;FORM ADDR OF OTH
;ELEMENT OF BASIC
;ARRAY AT FE-FF
CLC
LDA $77
ADC #8
STA ARRAY
LDA #0
ADC $78
STA ARRAY+1
LDY #0
LDA (ARRAY),Y
STA PCNT
;INCR Y TO 1ST VALUE
INY
INY
;SET TO HEAD LEFT
LDA #$FF
```

```
STA PINC
LDA #LSTRT
STA PBASE
;SET FIRST PRNT MASK
JSR MASK
.PAGE
;TURN ON PRINTER
LDA #FRON
STA PCR
JSR PAT23
BNE NEWLIN
JSR PAT23
BNE NEWLIN
;NO PRINTER - RETURN
JMP $COD1
NEWLIN LDA #<T1
;TIME FIRST DOT
STA T2L
LDA #>T1
STA T2H
JSR DE2
;DO FIRST DOT
JSR PRNT
;DO FIVE DOTS
JSR DOFIV
;DO TWO MORE
LDX #2
DP1 LDA #<T2
STA T2L
LDA #>T2
STA T2H
JSR DE2
JSR PRNT
DEX
BNE DP1
.PAGE
;FIVE MORE
JSR DOFIV
;CHANGE DIRECTION
;(AFTER NEXT DOT)
INC PINC
BEQ SRGHT
;HEAD LEFT
LDA #$FF
STA PINC
LDA #LSTRT
STA PBASE
BNE CONT
;HEAD RIGHT
SRGHT INC PINC
LDA #RSTRT
STA PBASE
```

```
CONT INY
INY
;IF Y=0, INC ARRAY+1
BNE CONT1
INC ARRAY+1
;DO LAST DOT ON LINE
CONT1 LDA #<T3
STA T2L
LDA #>T3
STA T2H
JSR DE2
JSR PRNT
;MORE LINES?
DEC PCNT
BEQ ENDP
JMP NEWLIN
;NO: KILL PRINTER
ENDP LDA #$E1
STA PCR
;RETURN TO BASIC
JMP $COD1
.PAGE
;
;SUBROUTINE DOFIV
;  DOES FIVE DOTS
DOFIV LDX #5
;WAIT FOR NEXT CHAR
;DOT POSITION
DO1 LDA IFR
AND #MSP12
BEQ DO1
LDA PCR
EOR #1
STA PCR
JSR PRNT
DEX
BNE DO1
RTS
;
;SUBROUTINE PRINT
;  DOES PRINTING
;**** DO NOT MESS
;**** WITH PRTIME
;**** THERMAL ELEMNT
;**** DAMAGE MAY
;**** RESULT!!!!!
;
;LOAD MASKS TO ELMNT
PRNT LDA PMSKH
ORA DRB
STA DRB
LDA PMSKL
```

317

```
STA DRAH                        BPL SM1
;START TIMER                    ;YES: CLEAR MASK
LDA #<PRTIME                    LDA #0
STA T2L                         STA PMSKH
LDA #>PRTIME                    SM2 LDA PMSKH
STA T2H                         AND #3
;WHILE WAITING, SET             STA PMSKH
; NEXT MASK                     RTS
JSR MASK                        PCNT .BYTE 0
JSR DE2                         PBASE .BYTE 0
;SHUT OFF ELEMENTS              PINC .BYTE 0
LDA DRB                         PCOL .BYTE 0
AND #$FC                        PMSKL .BYTE 0
STA DRB                         PMSKH ,BYTE 0
LDA #0                          .PAGE
STA DRAH                        LSTRT=140
RTS                             RSTRT=125
.PAGE                           ADDR=$B00
;                               T1=$1700
;SUBROUTINE MASK                T2=$600
;   SETS PRINT ELEMNT           T3=$700
MASK LDA #1                     PRON=$C1
STA PMSKL                       MSP12=2
LDA #0                          PCR=$A80C
STA PMSKH                       IFR=$A80D
;MASK SET TO                    DRB=$A800
; 00 0000 0001 BNRY             DRAH=$A801
;GET OLD STRT ELEMNT            T2L=$A808
LDA PBASE                       T2H=$A809
CLC                             DE2=$EC1B
;ADD INCR                       PAT23=$FFA0
ADC PINC                        PRTIME=$06A4
STA PBASE                       .END
;STORE STRT ELEMNT
;IN PCOL
SM1 STA PCOL
;PCOL=ARRAY(Y)?
CMP (ARRAY),Y
BEQ SM2
;NO: SHIFT MASK LEFT
ASL PMSKL
ROL PMSKH
;ADJUST PCOL
LDA PCOL
SEC
SBC #14
;ALL COLUMNS DONE?
```

The above program can be assembled with the AIM Assembler ROM or
entered from the listing beginning at address 0E00.

```
10 REM BSDL2
20 REM DATA LOGGER
30 REM SKELETON
40 DIM A%(100)
50 PRINT FRE(0)
60 END
5000 REM SAMPLE
5010 INPUT "SMPL TIME:";ST
5020 ST=(ST*1000-2)/2
5030 IF ST>32767 THEN ST=ST-65536
5040 INPUT "CYCLES:";CY
5050 POKE 4,0:POKE 5,14
5060 FOR I=1 TO CY
5070 X=USR(ST)
5080 NEXT I
5090 PRINT "TAKEN"
5100 RETURN
6000 REM DAC
6010 PRINT "SCALING"
6020 MX=A%(1):MN=MX
6030 FOR I=1 TO A%(0)
6040 IF A%(I)<MN THEN MN=A%(I)
6050 IF A%(I)>MX THEN MX=A%(I)
6060 NEXT I
6065 IF MN=MX THEN PRINT "ALL SAME":RETURN
6070 SC=200/(MX-MN)
6080 FOR I=1 TO A%(0)
6090 A%(I)=(A%(I)-MN)*SC+50
6100 NEXT I
6110 T=A%(A%(0)):A%(A%(0))=0
6115 PRINT "LOOK"
6120 POKE 4,67:POKE 5,14
6130 X=USR(400)
6131 A%(A%(0))=T
6140 SC=130/200
6150 FOR I=1 TO A%(0)
6160 A%(I)=(A%(I)-50)*SC
6170 NEXT I
6180 POKE 4,208
6190 X=USR(1000)
6195 PRINT MN;MX
6200 RETURN
```

```
          'DTLGR.ASM 30-JUL-81'
;J. AVERY
;
;ADAC COMBINES THE
;ADC AND DAC RTNS
;DEVELOPED IN CH 13
;AND REDUCES THE
;BASIC OVERHEAD BY
;PREFORMING THE VIA
;SET UP
;
;TO ADD N SETS OF
;VALUES TO THE FIRST
;I POINTS IN THE A%
;ARRAY:
;
;DIM A%(128)
;(MUST BE 1ST ARRAY
; AND 128 IS MAX
;
;POKE 4,0
;POKE 5,14
;A%(0)=I
;FOR J=1 TO N
;Z=USR(T)
;(T IS PERIOD IN
; USEC; >150)
;
;TO DISPLAY THE 1ST
;I POINTS OF A%()
;
;DIM A%()
;(SCALE 0=<A%()=<127
;A%(0)=I
;POKE 4,67
;POKE 5,14
;Z=USR(T)
;(T PERIOD IN USEC,
; T>40 WORKS BEST)
;
;TO STOP DISPLAY,
;PULSE CB2 LINE
;
;BOXCAR ROUTINE
;
;RETURNS ONE VALUE
;TAKE X USEC AFTER
;FALLING EDGE OF PB7
;
;POKE 4,157
;POKE 4,14

;V=USR(X)
;
;ADC PROGRAM
;
*=$0E00
;SETUP
AD JSR SETUP
;
;GET # OF POINTS
LDY #00
LDA $FD
TAX
;WAIT FOR SYNC PULSE
;ON CA2
LDA $A001
ADSY1 LDA $A00D
AND #01
BEQ ADSY1
;START CLOCK
;T1 FREE, OUTPT
;LATCH PORT A
LDA #$C1
STA $A00B
LDA $AD
STA $A004
LDA $AC
STA $A005
;CLEAR CA1
LDA $A001
;MOVE POINTER AND
;WAIT FOR DATA
;SYNC ON CA1
ADLP1 INY
ADSY2 LDA $A00D
AND #02
BEQ ADSY2
;TAKE DATA AND ADD
;TO PREVIOUS VALUES
CLC
LDA $A001
ADC ($FE),Y
STA ($FE),Y
DEY
LDA #00
ADC ($FE),Y
STA ($FE),Y
INY
INY
;CHECK FOR ALL DONE
DEX
BNE ADLP1

JMP $C0D1
;
;DAC PROGRAM
;
;SETUP
DA JSR SETUP
;START CLOCK
;USE T1 FREE, NO OUT
LDA #$40
STA $A00B
LDA $AD
STA $A004
LDA $AC
STA $A005
;GET # OF POINTS
DALP1 LDY #00
LDA $FD
TAX
;SYNC ON CLOCK
DASY LDA $A00D
AND #$40
BEQ DASY
;CLEAR CLOCK FLAG
LDA $A004
;MOVE POINTER AND
;DISPLAY DATA
DALP2 INY
LDA ($FE),Y
STA $A000
INY
;DONE?
DEX
BNE DASY
;CHECK FOR CB2
LDA $A00D
AND #08
BEQ DALP1
;DONE...
JMP $C0D1
;
;SETUP ROUTINE
;
;GET T INTO AD/AC
SETUP JSR $BEFE
;GET ADDRESS OF 1ST
;ARRAY INTO FE/FF
CLC
LDA $77
ADC #08
STA $FE
LDA #00
```

```
ADC $78                    ;WAIT FOR CONVERSION
STA $FF                    BXSY LDA $A00D
;SET POINTER TO            AND #02
;THE FIRST ELEMENT         BEQ BXSY
;IN THE DATA ARRAY         ;TAKE DATA
LDY #00                    LDA $A001
LDA ($FE),Y                ;PUT DATA IN A,Y
STA $FD                    TAY
INC $FE                    LDA #00
BNE ST1                    JMP $COD1
INC $FF                    .END
;SET VIA:
;PORT A INPUT
ST1 LDA #00
STA $A003
;PORT B OUTPUT
LDA #$FF
STA $A002
;CA1 -- NEG EDGE
;CA2: POS, ORA/IFRO
;    CLEAR
;CB1: NEG
;CB2: NEG, IFR3
;   CLEAR
LDA #$24
STA $A00C
;CLEAR FLAGS
LDA #$FF
STA $A00D
RTS
;
;BOXCAR ROUTINE
;
;DO SETUP
JSR SETUP
;SET CLOCK MODE
; ONE-SHOT, OUTPUT
LDA #$81
STA $A00B
;SET T1
LDA $AD
STA $A004
LDA $AC
STA $A005
;CLEAR CA FLAGS
LDA $A001
```

The above program can be assembled with the AIM Assembler ROM or entered from the listing beginning at address 0ED0.

The following listing was prepared using the AIM Monitor K command.
The data acquisition and plot routines are entered into memory using
the I command.  To save on tape, use the D (dump memory) command,
specifying ED0 as "from address", FFF as "to address", OUT=T, F=DTLGR.

```
K>*=E00                          0E64 C8  INY
/99                              0E65 B1  LDA (FE),Y
0E00 20  JSR 0E78                0E67 8D  STA A000
0E03 A0  LDY #00                 0E6A C8  INY
0E05 A5  LDA FD                  0E6B CA  DEX
0E07 AA  TAX                     0E6C D0  BNE 0E5A
0E08 AD  LDA A001                0E6E AD  LDA A00D
0E0B AD  LDA A00D                0E71 29  AND #08
0E0E 29  AND #01                 0E73 F0  BEQ 0E55
0E10 F0  BEQ 0E0B                0E75 4C  JMP C0D1
0E12 A9  LDA #C1                 0E78 20  JSR BEFE
0E14 8D  STA A00B                0E7B 18  CLC
0E17 A5  LDA AD                  0E7C A5  LDA 77
0E19 8D  STA A004                0E7E 69  ADC #08
0E1C A5  LDA AC                  0E80 85  STA FE
0E1E 8D  STA A005                0E82 A9  LDA #00
0E21 AD  LDA A001                0E84 65  ADC 78
0E24 C8  INY                     0E86 85  STA FF
0E25 AD  LDA A00D                0E88 A0  LDY #00
0E28 29  AND #02                 0E8A B1  LDA (FE),Y
0E2A F0  BEQ 0E25                0E8C 85  STA FD
0E2C 18  CLC                     0E8E E6  INC FE
0E2D AD  LDA A001                0E90 D0  BNE 0E94
0E30 71  ADC (FE),Y              0E92 E6  INC FF
0E32 91  STA (FE),Y              0E94 A9  LDA #00
0E34 88  DEY                     0E96 8D  STA A003
0E35 A9  LDA #00                 0E99 A9  LDA #FF
0E37 71  ADC (FE),Y              0E9B 8D  STA A002
0E39 91  STA (FE),Y              0E9E A9  LDA #24
0E3B C8  INY                     0EA0 8D  STA A00C
0E3C C8  INY                     0EA3 A9  LDA #FF
0E3D CA  DEX                     0EA5 8D  STA A00D
0E3E D0  BNE 0E24                0EA8 60  RTS
0E40 4C  JMP C0D1                0EA9 20  JSR 0E78
0E43 20  JSR 0E78                0EAC A9  LDA #81
0E46 A9  LDA #40                 0EAE 8D  STA A00B
0E48 8D  STA A00B                0EB1 A5  LDA AD
0E4B A5  LDA AD                  0EB3 8D  STA A004
0E4D 8D  STA A004                0EB6 A5  LDA AC
0E50 A5  LDA AC                  0EB8 8D  STA A005
0E52 8D  STA A005                0EBB AD  LDA A001
0E55 A0  LDY #00                 0EBE AD  LDA A00D
0E57 A5  LDA FD                  0EC1 29  AND #02
0E59 AA  TAX                     0EC3 F0  BEQ 0EBE
0E5A AD  LDA A00D                0EC5 AD  LDA A001
0E5D 29  AND #40                 0EC8 A8  TAY
0E5F F0  BEQ 0E5A                0EC9 A9  LDA #00
0E61 AD  LDA A004                0ECB 4C  JMP C0D1
```

Plot Routine

The following listing was prepared using the AIM Monitor K command.
The data acquisition and plot routines are entered into memory using
the I command.  To save on tape, use the D (dump memory) command,
specifying ED0 as "from address", FFF as "to address", OUT=T, F=DTLGR.

```
K>*=ED0                OF3F DO BNE OF49       OFB6 8D STA OFEF
/99                    OF41 EE INC OFED       OFB9 A9 LDA #00
  OED0 18 CLC          OF44 A9 LDA #7D
  OED1 A5 LDA 77       OF46 8D STA OFEC
  OED3 69 ADC #08      OF49 C8 INY            <K>*=FBB
  OED5 85 STA FE       OF4A C8 INY            /99
  OED7 A9 LDA #00      OF4B DO BNE OF4F        OFBB 8D STA OFF0
  OED9 65 ADC 78       OF4D E6 INC FF          OFBE AD LDA OFEC
  OEDB 85 STA FF       OF4F A9 LDA #00         OFC1 18 CLC
  OEDD AO LDY #00      OF51 8D STA A808        OFC2 6D ADC OFED
  OEDF B1 LDA (FE),Y   OF54 A9 LDA #07         OFC5 8D STA OFEC
  OEE1 8D STA OFEB     OF56 8D STA A809        OFC8 8D STA OFEE
  OEE4 C8 INY          OF59 20 JSR EC1B        OFCB D1 CMP (FE),Y
  OEE5 C8 INY          OF5C 20 JSR OF87        OFCD FO BEQ OFE2
  OEE6 A9 LDA #FF      OF5F CE DEC OFEB        OFCF OE ASL OFEF
  OEE8 8D STA OFED     OF62 FO BEQ OF67        OFD2 2E ROL OFF0
  OEEB A9 LDA #8C      OF64 4C JMP OF05        OFD5 AD LDA OFEE
  OEED 8D STA OFEC     OF67 A9 LDA #E1         OFD8 38 SEC
  OEFO 20 JSR OFB4     OF69 8D STA A80C        OFD9 E9 SBC #0E
  OEF3 A9 LDA #C1      OF6C 4C JMP COD1        OFDB 10 BPL OFC8
  OEF5 8D STA A80C     OF6F A2 LDX #05         OFDD A9 LDA #00
  OEF8 20 JSR FFAO     OF71 AD LDA A80D        OFDF 8D STA OFF0
  OEFB DO BNE OF05     OF74 29 AND #02         OFE2 AD LDA OFF0
  OEFD 20 JSR FFAO     OF76 FO BEQ OF71        OFE5 29 AND #03
  OF00 DO BNE OF05     OF78 AD LDA A80C        OFE7 8D STA OFF0
  OF02 4C JMP COD1     OF7B 49 EOR #01         OFEA 60 RTS
  OF05 A9 LDA #00      OF7D 8D STA A80C        OFEB 00 BRK
  OF07 8D STA A808     OF80 20 JSR OF87        OFEC 00 BRK
  OFOA A9 LDA #17      OF83 CA DEX             OFED 00 BRK
  OFOC 8D STA A809     OF84 DO BNE OF71        OFEE 00 BRK
  OFOF 20 JSR EC1B     OF86 60 RTS             OFEF 00 BRK
  OF12 20 JSR OF87     OF87 AD LDA OFF0        OFFO 00 BRK
  OF15 20 JSR OF6F     OF8A OD ORA A800        OFF1 3D AND A218,X
  OF18 A2 LDX #02      OF8D 8D STA A800        OFF4 D4 ???
  OF1A A9 LDA #00      OF90 AD LDA OFEF        OFF5 30 BMI 1069
  OF1C 8D STA A808     OF93 8D STA A801        OFF7 38 SEC
  OF1F A9 LDA #06      OF96 A9 LDA #A4         OFF8 91 STA (12),Y
  OF21 8D STA A809     OF98 8D STA A808        OFFA 20 JSR F4A7
  OF24 20 JSR EC1B     OF9B A9 LDA #06         OFFD 9B ???
  OF27 20 JSR OF87     OF9D 8D STA A809        OFFE FF ???
  OF2A CA DEX          OFAO 20 JSR OFB4        OFFF A8 TAY
  OF2B DO BNE OF1A     OFA3 20 JSR EC1B        1000 10 BPL 1012
  OF2D 20 JSR OF6F     OFA6 AD LDA A800        1002 10 BPL 1014
  OF30 EE INC OFED     OFA9 29 AND #FC         1004 10 BPL 1016
  OF33 FO BEQ OF41     OFAB 8D STA A800
  OF35 A9 LDA #FF      OFAE A9 LDA #00
  OF37 8D STA OFED     OFBO 8D STA A801
  OF3A A9 LDA #8C      OFB3 60 RTS
  OF3C 8D STA OFEC     OFB4 A9 LDA #01
```

```
5 REM FFT
10 DIMAZ(128),FR(128),FI(128)
15 N=128
16 AZ(O)=N
20 POKE 4,0:POKE 5,14
30 X=USR(500)
40 FORI=1TO128
50 FR(I)=AZ(I):FI(I)=0
60 NEXT I
63 PRINT  AZ(0)
65 Z=0:GOSUB 1400
70 GOSUB1000
80 Z=1:GOSUB1400
999 END
1000 PRINT"FFT ";N
1010 MR=0
1020 NN=N-1
1030 P=3.141592
1040 FORM=1TONN
1050 L=N
1060 L=L/2
1070 IF(MR+L)>NNTHEN1060
1080 MR=L+MR-INT(MR/L)*L
1090 IFMR=<MTHEN1160
1100 TR=FR(M+1)
1110 FR(M+1)=FR(MR+1)
1120 FR(MR+1)=TR
1130 TI=FI(M+1)
1140 FI(M+1)=FI(MR+1)
1150 FI(MR+1)=TI
1160 NEXT
1170 L=1
1180 IFL>=NTHENRETURN
1190 IS=2*L
1200 FORM=1TOL
1210 A=P*(1-M)/L
1220 WR=COS(A)
1230 WI=SIN(A)
1240 FORI=MTONSTEPIS
1250 J=I+L
1260 RJ=FR(J):IJ=FI(J):RI=FR(I):II=FI(I)
1270 TR=WR*RJ-WI*IJ
1280 TI=WR*IJ+WI*RJ
1290 FR(J)=RI-TR
1300 FI(J)=II-TI
1310 FR(I)=RI+TR
1320 FI(I)=II+TI
1330 NEXTI
1340 NEXTM
1350 L=IS
1360 GOTO1180
1400 PRINT"PLOT"
1405 MX=-1E10:MI=1E10
1410 FORI=1TON
```

```
1420  IFZ=0THENX=FR(I)
1430  IFZ><0THENX=SQR(FR(I)*FR(I)+FI(I)*FI(I))
1440  IFX>MXTHENMX=X
1450  IFX<MITHENMI=X
1460  NEXT
1470  S=100/(MX-MI)
1480  FORI=1TON
1490  IFZ=0THENA%(I)=(FR(I)-MI)*S
1500  IFZ><0THENA%(I)=(SQR(FR(I)*FR(I)+FI(I)*FI(I))-MI)*S
1510  NEXT
1520  POKE4,208
1530  POKE5,14
1540  X=USR(0)
1550  PRINTMI,MX
1560  RETURN
```

```
5 REM SMOTH
10 DIM A%(100),A(100),B(100),C(30),E(100)
15 POKE 4,208:POKE 5,14
20 X=RND(-1)
30 INPUT "FWHM";W
40 N1=2*W+10:IF N1>50 THEN N1=50
50 INPUT "SNR";S1
60 D=50/S1
70 C=-4*LOG(2)/(W*W)
80 FOR N=1 TO 2*N1
90 Y=50*EXP(C*(N-N1)*(N-N1))
100 A(N)=INT(Y):B(N)=A(N)
110 NEXT
115 F=1:L=2*N1
120 GOSUB 1000
130 FOR N=1 TO 2*N1
140 S2=0
150 FOR I=1 TO 12
160 S2=S2+RND(1)
170 NEXT I
180 A(N)=A(N)+(S2-6)*D:B(N)=A(N)
190 NEXT N
200 GOSUB 1000
300 FOR I=1 TO 100:B(N)=A(N):NEXT
310 INPUT "WIDTH OF SMOOTH";M
320 IF M>30 THEN 310
330 M=INT((M-1)/2)
340 REM DEPENDS ON SMOOTH
350 FOR S=-M TO M
360 REM DEPENDS ON SMOOTH
370 C(S+15)=N/D
380 PRINT S;C(S+15)
390 NEXT
500 F=M+1:L=2*N1-M
510 FOR I=F TO L
520 S2=0
530 FOR J=-M TO M
540 S2=S2+C(J+15)*B(I+J)
550 NEXT J
560 E(I)=S2
570 NEXT I
580 FOR I= F TO L:B(I)=E(I):NEXT
590 GOTO 200
1000 PRINT " "
1010 M1=A(1):M2=M1
1020 FOR N=F TO L
1030 IF B(N)>M1 THEN M1=B(N)
1040 IF B(N)<M2 THEN M2=B(N)
1050 NEXT N
1060 G=100/(M1-M2)
1070 FOR I=F TO L
1080 A%(I-F+1)=(B(I)-M2)*G
1090 NEXT I
1100 A%(0)=L-F+1
1110 X=USR(0)
1120 RETURN
```